Advances in Industrial Control

Springer
London
Berlin
Heidelberg
New York
Barcelona
Budapest
Hong Kong
Milan
Paris
Santa Clara
Singapore
Tokyo

Other titles published in this Series:

Intelligent Seam Tracking for Robotic Welding
Nitin Nayak and Asok Ray

Identification of Multivariable Industrial Process for Simulation, Diagnosis and Control
Yucai Zhu and Ton Backx

Nonlinear Process Control: Applications of Generic Model Control
Edited by Peter L. Lee

Microcomputer-Based Adaptive Control Applied to Thyristor-Driven D-C Motors
Ulrich Keuchel and Richard M. Stephan

Expert Aided Control System Design
Colin Tebbutt

Modeling and Advanced Control for Process Industries, Applications to Paper Making Processes
Ming Rao, Qijun Xia and Yiquan Ying

Robust Multivariable Flight Control
Richard J. Adams, James M. Buffington, Andrew G. Sparks and Siva S. Banda

Modelling and Simulation of Power Generation Plants
A.W. Ordys, A.W. Pike, M.A. Johnson, R.M. Katebi and M.J. Grimble

Model Predictive Control in the Process Industry
E.F. Camacho and C. Bordons

H_∞ Aerospace Control Design: A VSTOL Flight Application
R.A. Hyde

Neural Network Engineering in Dynamic Control Systems
Edited by Kenneth Hunt, George Irwin and Kevin Warwick

Neuro-Control and its Applications
Sigeru Omatu, Marzuki Khalid and Rubiyah Yusof

Energy Efficient Train Control
P.G. Howlett and P.J. Pudney

Hierarchical Power Systems Control: Its Value in a Changing Industry
Marija D. Ilic and Shell Liu

System Identification and Robust Control
Steen Tøffner-Clausen

Genetic Algorithms for Control and Signal Processing
K.F. Man, K.S. Tang, S. Kwong and W.A. Halang

E.F. Camacho, M. Berenguel and F.R. Rubio

Advanced Control of Solar Plants

With 136 Figures

Springer

E.F. Camacho • M. Berenguel • F.R. Rubio
Departamento de Ingeniería de Sistemas y Automática
Escuela Superior de Ingenieros Industriales, Universidad de Sevilla
Avenida Reina Mercedes s/n, 41012 Sevilla, Spain

ISBN-13: 978-1-4471-1249-5 e-ISBN-13: 978-1-4471-0981-5
DOI: 10.1007/978-1-4471-0981-5

British Library Cataloguing in Publication Data
Camacho, E.F.
Advanced control of solar plants. - (Advances in industrial control)
1.Solar collectors 2.Solar power plants - Automatic control
I.Title II.Berenguel, Manuel III.Rubio, Francisco R.
621.4'72

Library of Congress Cataloging-in-Publication Data
Camacho, E. F.
Advanced control of solar plants / E.F. Camacho, M. Berenguel, and F.R. Rubio.
p. cm. - - (Advances in industrial control)
Includes bibliographical references (p.) and index.

1. Solar power plants- -Spain- -Almería (Province). 2. Solar thermal energy- -Spain- -Almería (Province). 3. Electric power systems- -Spain- -Almería (Province)- -Control. I. Berenguel, M. (Manuel), 1968- II. Rubio, F. R. (Francisco R.), 1954-
III. Title. IV. Series.
TK1056.C35 1997 97-4493
621.31'243- -dc21 CIP

Softcover reprint of the hardcover 1st edition 1997

Typesetting: Camera ready by authors

69/3830-543210 Printed on acid-free paper

SERIES EDITORS AND ADVISORY BOARD

To Janet
E.F.C.

To Mª Paz, Isa, Víctor and Juan Fco
M.B.

To María
F.R.R.

SERIES EDITORS' FOREWORD

The series *Advances in Industrial Control* aims to report and encourage technology transfer in control engineering. The rapid development of control technology impacts all areas of the control discipline. New theory, new controllers, actuators, sensors, new industrial processes, computer methods, new applications, new philosophies..., new challenges. Much of this development work resides in industrial reports, feasibility study papers and the reports of advanced collaborative projects. The series offers an opportunity for researchers to present an extended exposition of such new work in all aspects of industrial control for wider and rapid dissemination.

This volume by Professor Eduardo F. Camacho and his colleagues Manuel Berenguel and Francisco R. Rubio is an exemplar of what an *Advances in Industrial Control* monograph should be. In it the control of a thermal solar facility is used to study the performance obtainable from an interesting range of control algorithms. These methods range from the conventional PID controller, through to model-based predictive and robust optimal control methods and finishing with two fuzzy logic based control techniques. The scientific methodology applied is modelling, simulation and plant implementation. In the last chapter, a rigorous approach for a comparative study is described involving a careful selection of performance metrics. The text is rich in relevant up-to-date source material, and contains many thought-provoking comments. The presentation is well-balanced, impartial and very readable.

The volume should appeal to all who have an interest in reading about how advanced control performs in practice. Each method is described in short form making it a useful compendium of the essentials for many of the techniques available. Most invaluable is the practical insight described in the volume of the authors' real applications experience obtained at the Plataforma Solar de Almeria in Spain.

M.J. Grimble and M.A. Johnson
Industrial Control Centre
Glasgow, Scotland, UK

lowed by a wide range of readers, as only basic knowledge of control theory and sampled data systems is required.

The book is organized as follows: Chapter 1 gives a brief introduction to the solar plant and the control techniques used in the text. Chapter 2 contains a description of distributed solar collector fields. Dynamic models, both linear and nonlinear are described. Chapter 3 is devoted to presenting the basic control scheme and the results obtained when controlling the plant with PID type of controllers. In Chapter 4 an adaptive PI control scheme is discussed. Based on the dynamical characterization of the field, some typical advanced control strategies are presented: Model based predictive control strategies (Chapter 5), frequency based control and robust optimal control (Chapter 6) and heuristic fuzzy logic control (Chapter 7). Chapter 8 is devoted to presenting a summary of the features of the different control schema and some concluding remarks.

The text is composed of material collected from articles written by the authors, technical reports and lectures given to graduate students.

Acknowledgments

The authors would like to thank a number of people and institutions who have made this book possible. Firstly we thank Janet Buckley who translated part of the book from our native language to English and corrected and polished the style of the rest. Our thanks to Javier Aracil who introduced us to the exciting world of Control and to many other colleagues and friends from our Department, especially F. Gordillo, who helped us to correct the manuscript by discussing some of the ideas appearing in the book.

Most of the material included in the book is the result of research work funded by the CICYT[1], CIEMAT, the European Commission[2] and IBERDROLA[3]. We gratefully acknowledge these institutions for their support.

The experiments described in the book could not have been carried out without the help of the Plataforma Solar de Almería (PSA) and their personnel, especially Pedro Balsa, Eduardo Zarza, Diego Martínez, Javier León

[1] Grants TAP-93-0408 and TAP-95-0370.

[2] Project *Enhancement and Development of Industrial Applications of Solar Energy Technologies*, supported by EC Program *Human Capital and Mobility - Large Installations Program* and promoted by CIEMAT - Plataforma Solar de Almería, Spain.

[3] Programa de Ayudas a la Investigación Científica y al Desarrollo Tecnológico, Proyecto 036/96.

and Jaime Aranda. We thank them and the rest of the personnel at the PSA who helped us to carry out the tests.

Finally the authors thank their families for their support, patience and understanding of family time lost during the writing of the book.

Sevilla, April 1997

Eduardo F. Camacho, Manuel Berenguel and Francisco R. Rubio

Address:	Departamento de Ingeniería de Sistemas y Automática Escuela Superior de Ingenieros. Universidad de Sevilla Avenida Reina Mercedes s/n, 41012, Sevilla, SPAIN
E-mail:	`[eduardo,beren,rubio]@cartuja.us.es`
Home page:	`http://www.esi.us.es/~[eduardo,beren,rubio]`
Fax:	`34-5-4556849`
Telf:	`34-5-4556865/67/76`

Software, comments and questions

A simulation software package for the ACUREX distributed solar collector field (nonlinear distributed parameter model for simulation purposes described in the book) can be obtained by contacting the authors or by accessing the author's home pages. Please, send errors, comments and questions to the authors.

CONTENTS

LIST OF TABLES

GLOSSARY

Acronyms

AI	*Artificial Intelligence*
AIC	*Akaike's Information Criterion*
ARIMA	*Auto Regressive Integrated Moving Average*
ARMA	*Auto Regressive Moving Average*
CARIMA	*Controlled Auto Regressive Integrated Moving Average*
CARMA	*Controlled Auto Regressive Moving Average*
CE	Control Effort
CRHPC	*Constrained Receding Horizon Predictive Control*
CRS	*Central Receiver System*
DAS	*Data Acquisition System*
DCS	*Distributed Collector System*
DMC	*Dynamic Matrix Control*
DSCF	*Distributed Solar Collector Field*
EHAC	*Extended Horizon Adaptive Control*
EPSAC	*Extended Prediction Self Adaptive Control*
FAM	*Fuzzy Associative Memory*
FF	*Feedforward*
FIR	*Finite Impulse Response*
FLC	*Fuzzy Logic Control*
GMV	*Generalized Minimum Variance*
GPC	*Generalized Predictive Control*
IFPIC	*Incremental Fuzzy PI Control*
IMC	*Internal Model Control*
I/O	*Input / Output*
ITAE	*Integral Time Absolute Error*
KBS	*Knowledge Based System*
LCP	*Linear Complementary Problem*
LQ	*Linear Quadratic*
LQG	*Linear Quadratic Gaussian*
LS	*Least Squares*
LTI	*Linear Time Invariant*
LTR	*Loop Transfer Recovery*

MAC	*Model-based Algorithmic Control*
MIMO	*Multi Input - Multi Output*
MIN-MAX	*Minimum-Maximum*
MPC	*Model-based Predictive Control*
MPHC	*Model Predictive Heuristic Control*
MRAC	*Model Reference Adaptive Control*
NI	*Numeric Inference*
NN	*Neural Networks*
NNLS	*Non Negative Least Squares*
PC	*Personal Computer*
PFC	*Predictive Functional Control*
PID	*Proportional Integral Derivative*
PBSC	*Predictor Based Self-tuning Control*
PRBS	*Pseudo Random Binary Sequence*
PSA	*Plataforma Solar de Almería*
QP	*Quadratic Programming*
RLS	*Recursive Least Squares*
SIORHC	*Stabilizing Input/Output Receding Horizon Control*
SGPC	*Stable GPC*
SISO	*Single Input / Single Output*
SSPS	*Small Solar Power Systems*
STC	*Self Tuning Control*
TEP	*Thermosolar Electrical Plants*
TP	*Triangular Partition*
TPE	*Triangular-Partition with Evenly spaced midpoints*
UBB	*Unknown But Bounded*
Z-N	*Ziegler-Nichols*

Model parameters and variables

d	Input-output delay (integer)
$R(t), r(t)$	Control loop reference or set point at instant t
$U(t), u(t)$	Plant input at instant t
$Y(t), y(t)$	Plant output at instant t
$Y_m(t), y_m(t)$	Plant model output at instant t
$E(t), e(t)$	Tracking error at instant t

Symbols

$abs()$	Absolute value of a number
$arg()$	Argument
$det()$	Matrix determinant
$tr()$	Matrix trace
$\mathbb{R}$	Set of real numbers
$sin()$	Sine function
$cos()$	Cosine function
$exp()$	Exponential function
$log()$	Natural logarithm function
$min()$	Minimum of a set
$max()$	Maximum of a set
s	Complex variable used in Laplace Transform
z^{-1}	Backward shift operator
z	Forward shift operator and complex variable used in Z-Transform
$\forall$	For all
$\Vert \cdot \Vert$	$C^{n \times m}$ norm
$\Vert \cdot \Vert_2$	$L_2^{n \times m}$ norm
$\Vert \cdot \Vert_\infty$	∞ norm
$\doteq$	Definition
$(\mathbf{M})_{ij}$	Element ij of matrix $\mathbf{M}$
$(\mathbf{v})_{ij}$	i^{th} element of vector $\mathbf{v}$
$(\cdot)^T$	Transpose of $(\cdot)$
$\mathbf{I}$	Identity matrix of appropriate dimension
$E[\cdot]$	Expectation operator
$\hat{}$	Expected value
$\hat{x}(t+j\vert t)$	Expected value of $x(t+j)$ with available information at instant t

Frequently-used quantities and units

Length	metre	m
Area	square metre	m^2
Volume	cubic metre	m^3
Time	seconds	s
	hours	h
Frequency	radians per second	rad/s
Density	kilogramme per cubic metre	Kg/m^3
Solar radiation	watts per square metre	W/m^2
Flow	cubic metre per second	m^3/s
	litre per second	l/s
Temperature	degrees (Celsius)	^oC
Angle	degrees	^{o}
	radian	rad

Greek characters

γ	Multiplicative constant for closed loop pole location	
Γ	Diophantine equation polynomial	
δ	Weighting sequence in GPC	
δ_l	Declination	^{o}
δ_2	Hourly angle	^{o}
Δ	Variation in a magnitude	
Δ_c	Return difference function	
$\triangle$	$1 - z^{-1}$ operator	
ϵ	Zero mean noise and estimation error	
ζ	Number slightly greater than one	
$\boldsymbol{\theta}$	Estimated parameters vector	
τ_r	Residence time	s
λ	Weighting sequence in GPC	
$\mu(x)$	Membership function of a fuzzy set	
μ	Dynamic viscosity	Pa.s
π	PI	
Π	Product	
$\Pi(z)$	Monic polynomial with roots of multiplicity one inside the unit circle	
ρ	Thermal oil density	Kg/m^3
ρ_0	Bound of an analytic function	
Σ	Addition	
Υ	Diophantine equation polynomial	
τ	Characteristic time constant	s
τ_d	Time delay	s
ϕ	Incidence angle	^{o}
$\boldsymbol{\phi}_k$	Last input-output vector for identification	
$\boldsymbol{\Phi}$	State transition matrix	
φ	Last input-output vector for robust identification	
χ	Predictable disturbances	
ψ	Stochastic disturbances	
$\boldsymbol{\Psi}$	Input-output signal matrix for robust control	
ω	Frequency	rad/s
Ω	Set of points equally spaced around the unitary circle	
Ω_s	Set of odd interpolating points	

CHAPTER 1
INTRODUCTION

The Control Theory for linear processes has for some time been considered a well established scientific discipline with powerful techniques for analyzing and designing controllers. The main problems in process control when applying the Linear Control Theory are caused by the fact that:

1. A linear mathematical model of the plant is needed and finding one is not a trivial problem in many cases.
2. Mathematical models of real processes cannot take all aspect of reality into account. Simplifying assumptions have to be made and models are only approximations of reality.
3. Most processes are nonlinear, having nonlinear dynamics and nonlinearities caused by actuators that have a limited range of action and a limited slew rate, as in the case of control valves, which are limited by fully closed and fully open positions and a maximum slew rate. Constructive and/or safety reasons, as well as sensor ranges, cause limits in process variables, as in the case of tank levels, pipe flows and pressures in deposits.
4. Because of changing environmental conditions, such as ambient temperature, humidity etc., most processes are not time invariant.

These problems have been extensively treated in literature and some new disciplines have appeared to address them. Some of the disciplines have evolved around the Linear Systems Control Community, as is the case of Robust or Adaptive Control while other disciplines have developed around the Artificial Intelligence Control Community, as is the case of Expert, Fuzzy, or Neural Control.

In Robust Control the process is usually modelled by a linear model and some of the problems mentioned above are treated by considering uncertainties about the model. The main assumption in most cases is that the underlying process is linear. In Adaptive Control the main idea is that by an appropriate adaptation mechanism, the controller and/or model of the process, which in most cases is linear, will cope with unknown, changing and possibly nonlinear dynamics. Nonlinear control strategies, normally based on an exact cancelation of the nonlinear dynamics have to be used for nonlinear processes such as robots. The uncertainties in the dynamic parameters of

Table 1.1. Seborg's classification of process control strategies according to the degree of use in industry

Category I:	**Conventional strategies**
	- Manual control - PID control - Ratio control - Cascade control - Feedforward control
Category II:	**Advanced control: classical techniques**
	- Gain scheduling - Time delay compensation - Decoupling control - Selective/override controllers
Category III:	**Advanced control: widely used techniques**
	- Model predictive control (MPC) - Statistical quality control - Internal model control (IMC) - Adaptive control
Category IV:	**Advanced control: newer techniques with some industrial applications**
	- Optimal control (LQG) - Expert systems - Nonlinear control - Neurocontrollers - Fuzzy control
Category V:	**Advanced control: proposed strategies with few industrial applications**
	- H_∞ control - μ-synthesis

Table 1.2. Control strategies treated in the book

Methodology	Controller
Model-based predictive control	Adaptive GPC control with reduced linear models
	Robust adaptive GPC control with reduced linear models and bounded uncertainties
	Gain scheduling GPC control based on extended linear models
	GPC control with nonlinear prediction of the system free response
Robust optimal control and frequency domain control	Frequency domain adaptive IMC
	Robust control LQG/LTR
Fuzzy logic control	Incremental fuzzy PI control (IFPIC)
	Fuzzy logic control (FLC)

taking into account the similarities of the control methodology needed, as shown in Table 1.2.

The rest of this chapter is dedicated to introducing the control techniques used in the rest of the book. That is, modelling and simulation, adaptive control, model predictive control, robust control and fuzzy control.

1.3 Modelling and Identification

According to Ljung [75], modelling means that all basic physical laws of the object being characterized are united and organized in a manner, (the model), that is useful for simulation, analysis or synthesis. A first step in developing a model is to determine the process variables that are relevant in the behaviour of the plant. A second step is the analysis of the way in which the variables are dynamically related. Two types of approaches can be distinguished for establishing these relationships: one is based on the physical laws and the other based on experiments carried out on the process which is considered to be a black box. In this second case, the black box represents a family of models (usually linear models) whose parameters do not have a physical meaning. The objective is to find a set of parameters that make the model behaviour fit to the observed data. The possibility of combining both approaches is also used and the models developed are known as grey box models.

The first modelling approaches in control were done in the frequency domain. The trend then changed to parametric identification for some time and a renewed interest in process characterization in the frequency domain has emerged in connection with frequency based adaptive control methods and robust control.

There are two main reasons making model identification of real processes difficult:

- The observed data cannot be reproduced exactly because of signals that cannot be measured and affect process dynamics and
- The observed data do not cover all possible situations in which the model is needed.

According to this, all models are subjected to two types of errors:

- Random errors: due to the fact that observations cannot be reproduced exactly.
- Systematic errors: due to the simplicity of the model structure chosen.

System identification is, therefore, an experimental approximation to process modelling and the following steps can be distinguished in this process:

- Planning the experiments to be carried out.
- Selecting the model structure.
- Estimating the model parameters.
- Validating the model.

1.4 Adaptive Control

The main idea behind adaptive control is to modify the controller when process dynamic changes. It can be said that adaptive control is a special kind of nonlinear control where the state variables can be separated into two groups moving in two different time scales. The state variables which change more rapidly, correspond to the process variables (internal loop), while the state components which change more slowly, correspond to the estimated process (or controller) parameters.

Since Aström and Wittenmark [6] first introduced self-tuning control, many works have appeared in literature describing new methodological developments and applications. Although some of these applications come from industry and there are some commercial adaptive controllers on the market, the technique has not made it to industry as expected, and even many controllers with adaptive capabilities are working as fixed controllers. The main reason for this slow progression of adaptive control in industry is that adaptive control, as other advanced control techniques, requires more knowledge from process operators as it has more tuning parameters and more concepts to be understood. Another fundamental problem is that when applied to a real process the two time scales are not as far apart as they should be and this may be a problem for parameter identification and for assessing stability. It was shown at the beginning of the eighties that adaptive control systems could be unstabilized by small external perturbations or small modelling errors. Robustness studies of adaptive control systems are, therefore, of paramount importance. It is necessary to know the frequency response of the process around the desired crossover frequency in order to assess how parameter changes can affect the closed loop behaviour. Other important issues that should be taken into account when designing an adaptive control are:

- Using all available information about the process.
- Establishing realistic specifications for the closed loop.
- Getting a robust identification mechanism.
- Dealing with high frequency uncertainties.
- Starting up the adaptation process.

Rational transfer functions are the most widely used process models in most works dealing with adaptive control. The advantage of using this type of model is that the number of process parameters to be estimated is small and identification is not difficult. The following problems can, however, be mentioned when this approach is used:

- Parameter estimation is very sensitive to unmodelled dynamics.
- The parameters of a rational transfer function can change significantly without an equivalent change of the process dynamics as is the case with poles and zeros closely located.
- Information about the orders, dead time and time scales is needed.

Some of these difficulties can be addressed by using a frequency response process model. Adaptive techniques based on the frequency response have been proposed in literature [11], [53] and [56]. Chapter 5 shows an adaptive control scheme based on these techniques.

1.4.1 Adaptive Control Structures

Adaptive controllers have traditionally been classified into one of the following families: model reference adaptive controllers (MRAC) and self-tuning controllers (STC).

Model Reference Adaptive Controllers (MRAC)

In this type of controller, the specifications are given by a reference model that indicates how the process output $Y(t)$ should behave ideally $Y_m(t)$ for a particular reference signal $R(t)$.

The control system can be considered to be composed of two loops: an inner loop formed by the plant and the controller and an external loop which changes the controller parameters in such a way that the process follows the model. The problem is to determine a stable adjusting mechanism that takes the error to zero.

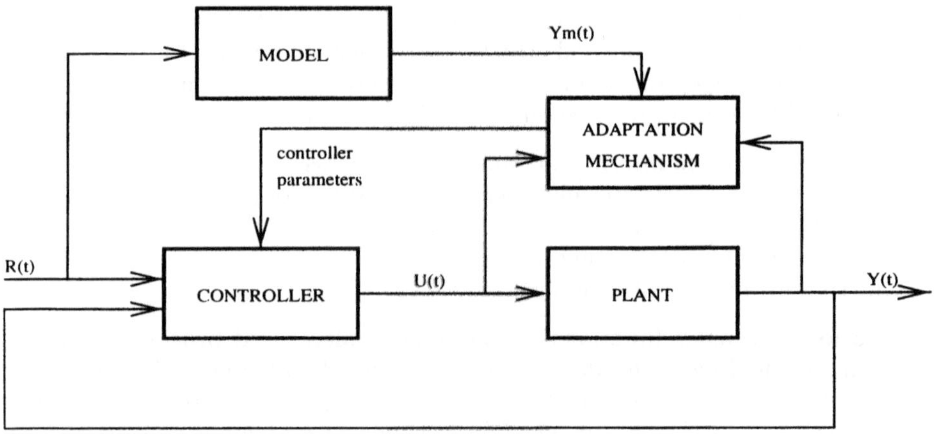

Fig. 1.1. Model reference adaptive control scheme (MRAC)

Self Tuning Controllers (STCs)

These types of controllers adapt to changes in the process dynamics as shown in Figure 1.2. The identifier block is continuously updating a model of the process which is used by the design block in order to set the controller parameters to values which are suitable for the process estimated model. In adaptive control, the estimation of the parameters is carried out in real time. Because of this, recursive mechanisms are normally used and special attention has to be given to the convergence of the recursive procedure as system stability depends on this. Different identification mechanisms are described in literature, but the one most used is the Recursive Least Squares (RLS) one due to its simplicity, stochastic properties and convergence characteristics. This mechanism is the one used in the adaptive control structures described in the book. Some techniques, such as the one based on the UDU factorization [16] or the one described in [87] based on a solution of a *Riccati equation*, can be used in order to obtain solutions which are numerically more robust than the original version of the algorithm.

The control parameters are updated in different ways depending on the design method chosen (minimum variance, pole placement, etc.). Notice that, while in MRAC the controller parameters are directly obtained, in STC the controller parameters are not updated by the recursive algorithm. This is referred to in literature as indirect adaptation. Sometimes an STC can work as a direct adaptive controller when the problem can be re-parameterized in such a way that the control parameters are directly estimated by the RLS algorithm. The control structure is composed of two loops: the inner loop that should evolve in a fast time scale corresponding to process variables,

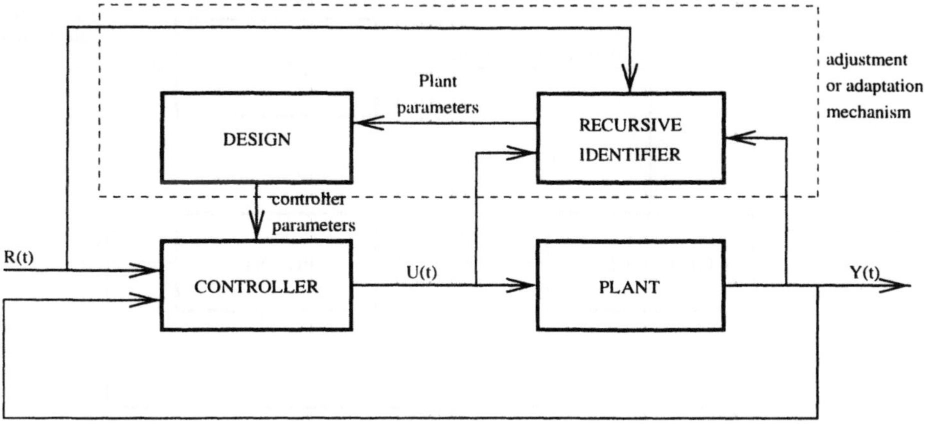

Fig. 1.2. Self-tuning control scheme (STC)

and the outer loop corresponding to model and controller parameters that should evolve in a much slower time scale.

Self-tuning control approximations

Some controllers have the capacity of adapting to changes in process dynamics but are not considered to be proper adaptive controllers. This is the case of *gain scheduling* controllers, where process dynamics can be associated to the value of some process variables that can be measured. This happens in processes which have their dynamics related to the operating point or to environmental conditions, as is the case of aircrafts which have dynamic characteristics which depend greatly on altitude and velocity. If the dynamic characteristics of the process can be inferred from measurable variables, the controller parameters can be computed from these variables as indicated in Figure 1.3. Notice that only the inner loop appears in this control structure and the parameter updating can be considered as a sort of feedforward which changes controller gains. These types of controllers have proved to be very useful in many situations, although they have certain disadvantages such as:

1. They cannot cope with dynamic changes due to wear and tear of process components.
2. The controller parameters have to be determined for different operating conditions covering all possible operating stages, and this requires a long commissioning time.
3. There are not many results results concerning the stability and robustness of these types of controllers (some recent results can for instance be found in [112]).

Another case of controllers that can change their parameters to adapt to process changes but are not considered as adaptive controllers are the

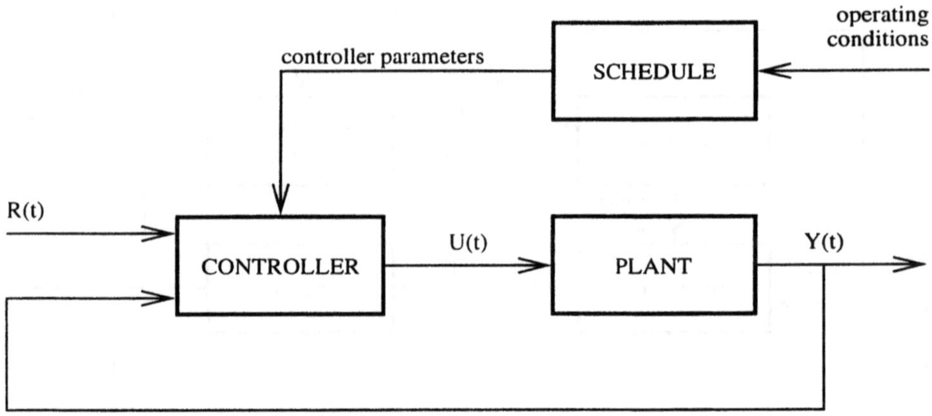

Fig. 1.3. Gain scheduling controller

auto-tuning ones. These types of controllers normally adapt their parameters under operator demand. They work in two phases:

- An experimental phase when a test signal is injected into the process, usually in an open loop configuration, and
- A parameter tuning phase. Process dynamics are inferred from the response to the test signal (usually a step or impulse) and the controller parameters (usually PIDs) are tuned by simple heuristic rules.

1.5 Model-based Predictive Control (MPC)

Predictive Control is made up of a very wide field of control methodologies which are developed using common ideas. Various disciplines are included such as optimal control, stochastic control, dead time process control, multivariable control or constrained control. A summary of the development of predictive control techniques can be found in [24].

Predictive control basically started to be developed in the sixties when various articles appeared reflecting the interest arising in the industry for model-based predictive control (MPC). This was fundamentally due to the simplicity of the algorithm and to the use of explicit dynamic models of the process (impulse response models, step response models, etc.). Using these models the effect of future control actions on the output is predicted, these are determined by minimizing the predicted error subjected to operational restrictions. Model-based Predictive Heuristic Control (MPHC), Model-based Algorithmic Control (MAC), both developed by Richalet *et al.* [98], [97] and Dynamic Matrix Control (DMC), developed by Cutler and Ramaker [38] can be mentioned amongst the most used methods in this context.

At the same time this was happening, another working line in connection with adaptive control ideas was appearing. The strategies developed were mainly based on formulations using input/output (I/O) models of the process.

Prediction Self-Adaptive Control by Peterka [92], Extended Horizon Adaptive Control (EHAC) by Ydstie [124], Extended Prediction Self-Adaptive Control (EPSAC) by De Keyser *et al.* [39], Generalized Minimum Variance control (GMV) by Clarke and Gawthrop [33], Predictive Functional Control (PFC) by Richalet *et al.* [96] and Generalized Predictive Control (GPC) by Clarke *et al.* [35] [36] can be mentioned in this context. Generalized predictive control (GPC) has become one of the most popular MPC methods.

Important works related to the stability and robustness of MPC are Constrained Receding Horizon Predictive Control (CRHPC) by Clarke and Scattolini [37], Stable Input/Output Receding Horizon Control (SIORHC) by Mosca *et al.* [88] and *Stable* GPC (SGPC) by Kouvaritakis *et al.* [63]. The MPC formulation in the state space can be found in [85]. In the field of robust design the works of Campo and Morari [28] and Allwright [2] must be mentioned; these implicitly take into account uncertainties in the design of controllers.

Model-based Predictive Control (MPC), is not a specific control strategy but more of a very wide range of control methods developed around certain common ideas. These design methods lead to linear controllers which have practically the same structure and present adequate degrees of freedom. The ideas appearing in greater or lesser degree in all the predictive control family are basically:

- Explicit use of a model to predict the process output at future time instants (horizon).
- Calculation of a control sequence minimizing a certain objective function.
- Receding strategy, so that at each instant the horizon is shifted towards the future, which involves the application of the first control signal of the sequence calculated at each step.

The various MPC algorithms only differ amongst themselves in the model used to represent the process and the noise and the cost function to be minimized. This type of control is of an open nature within which many works have been developed, being widely accepted by the academic world and by industry. There are many applications of predictive control successfully in use at the present time, not only in the process industry but also applications to the control of a diversity of processes ([31], [95]).

MPC presents a series of advantages over other methods, amongst which stand out:

- It is particularly attractive to staff with only a limited knowledge of control, because the concepts are very intuitive and at the same time the tuning is relatively easy.

- It can be used to control a great variety of processes, from those with relatively simple dynamics to other more complex ones, including systems with long delays, non-minimum phase or unstable ones.
- The multivariable case can easily be dealt with.
- It intrinsically has compensation for dead times.
- It introduces feedforward control in a natural way to compensate for measurable disturbances.
- The resulting controller is an easy to implement linear control law.
- Its extension to the treatment of constraints is conceptually simple and these can be included systematically during the design process.
- It is very useful when future references (robotics or batch processes) are known.
- It is a totally open methodology based on certain basic principles which allow for future extensions.

However, it also has its drawbacks. The greatest drawback is the need for an appropriate model of the process to be available. The design algorithm is based on a prior knowledge of the model and it is independent of it, but it is obvious that the benefits obtained will depend on the discrepancies existing between the real process and the model used.

In practice, MPC has proved to be a reasonable strategy for industrial control, in spite of the original lack of theoretical results at some crucial points such as stability or robustness.

MPC Structure

The methodology of all the controllers belonging to the MPC family is characterized by the following strategy, represented in Figure 1.4:

1. The future outputs for a determined horizon N, called the prediction horizon, are predicted at each instant t using the process model. These predicted outputs $y(t+k \mid t)$ [2] for $k = 1...N$ depend on the known values up to instant t (past inputs and outputs) and on the future control signals $u(t+k \mid t)$, $k = 0 \ldots N-1$, which are those to be sent to the system and to be calculated.
2. The set of future control signals is calculated by optimizing a determined criterion in order to keep the process as close as possible to the reference trajectory $w(t+k)$ (which can be the set point itself or a smooth approximation of it). This criterion usually takes the form of a quadratic function of the errors between the predicted output signal and the predicted reference trajectory. The control effort is included in the objective function in most cases. An explicit solution can be obtained if the criterion is quadratic, the model is linear and there are no constraints; otherwise an

[2] The notation indicates the value of the variable at the instant $t+k$ calculated at instant t.

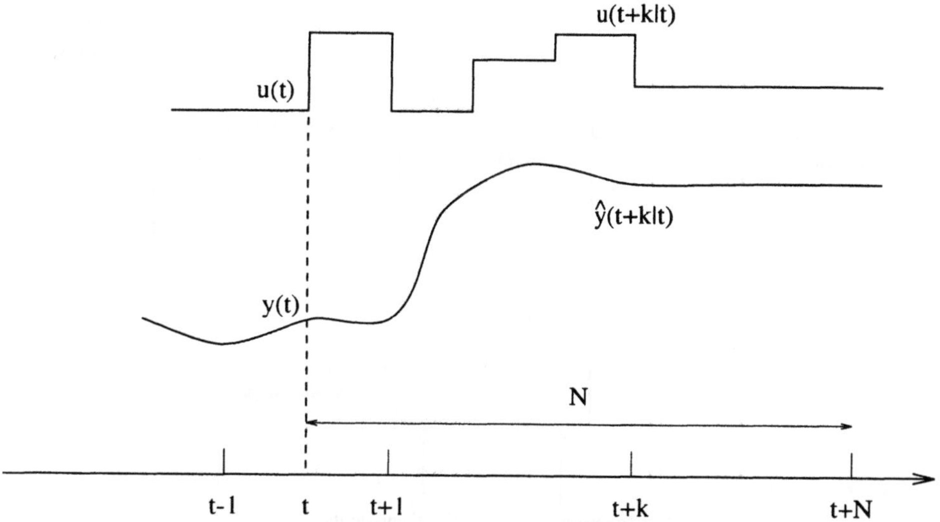

Fig. 1.4. MPC strategy

iterative optimization method has to be used. Some assumptions about the structure of the future control law are also made in some cases, such as that it will be constant from a given instant.

3. The control signal $u(t \mid t)$ is sent to the process while the next control signals calculated are rejected, because at the next sampling instant $y(t+1)$ is already known and step 1 is repeated with this new value and all the sequences are brought up to date. Thus the $u(t+1 \mid t+1)$ is calculated (which in principle will be different to the $u(t+1 \mid t)$ because of the new information available) using the receding horizon concept.

In order to implement this strategy, the basic structure shown in Figure 1.5 is used. A model is used to predict the future plant outputs, based on past and current values and on the proposed optimal future control actions. These actions are calculated by the optimizer taking into account the cost function (where the future tracking error is considered) as well as the constraints.

The process model plays, in consequence, a decisive role in the controller. The chosen model must be capable of capturing the process dynamics so as to precisely predict the future outputs as well as being simple to implement and to understand. As MPC is not a unique technique but a set of different methodologies, there are many types of models used in various formulations [24].

1.5.1 Nonlinear MPC control techniques

Because the majority of controlled processes have inherently nonlinear behaviour, the classical PID controllers, together with many advanced control

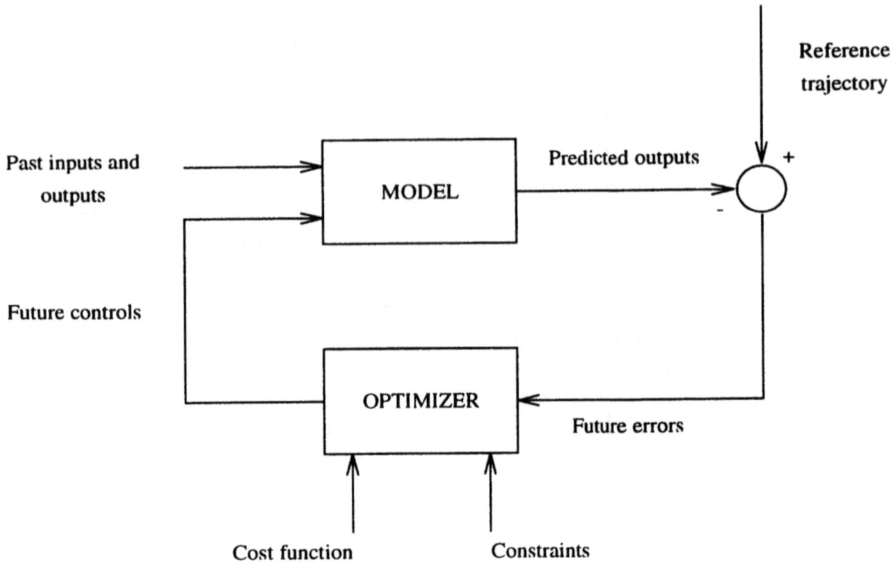

Fig. 1.5. Basic structure of MPC

structures based on linear models, should be sintonized conservatively in order to guarantee the stability of the controlled system under the entire range of possible operational conditions. Poor behaviour may be produced at critical points of the operation. In consequence, there are incentives to develop control strategies based on nonlinear process models, especially if said models can be obtained from physical principles of evolution of the characteristic process variables. This is a very active field of research, as is shown by recent publications [54], [64], etc.

The most recent techniques tend to be solutions to a specific problem (as is the case shown in this book) or they can be based on an approximation with a gain scheduling approach, where the control signals vary according to changes in the process. Numerous control strategies based on models which include artificial neural networks have recently appeared.

Also to be found in the literature are extensions of the MPC techniques to situations in which the model is nonlinear [15]. In this case a nonlinear programming problem must be solved in real time at every sampling period instead of the quadratic linear problem typical of standard MPC. The main difficulties of these methods are:

1. The theoretical analysis of properties of the closed loop such as stability and robustness is very complicated because of the appearance of nonlinear models in the formulation. Application or simulation examples are, therefore, usually used which do not guarantee the generality of the results or are not representative of them.

2. If the solution of solving a nonlinear programming problem at each sampling period is adopted, it is difficult to guarantee the convergence of the algorithm in an adequate lapse of time.

1.6 Robust control, frequency domain control and optimal control

As indicated in [100] the incorporation of a feedback structure in a control system aims, amongst other objectives, at:

1. Stabilizing the system;
2. Reducing the system sensibility to noise;
3. Specifying the transitory behaviour of the closed loop system (in time and/or in frequency domains);
4. Improving the system robustness to variations in the parameters of the plant.

There are many techniques for designing feedback systems with a high degree of robustness. Two major types of techniques can be distinguished: the classical frequency methods and the modern ones based on state variables.

The majority of the analysis and design methods for feedback systems presupposes that there exists a sufficiently exact model of the process to be controlled. In this model both the structure and the parameters are fixed. Some of the methods draw conclusions about the effect of the possible uncertainties but the treatment and rejection of these uncertainties is not their main function.

A more realistic approach for feedback control should take into account the model imprecisions, which are the main reason for using feedback. Another way of tackling the problem is to use principles and methods which allow the discrepancies between the model and the real process to be explicitly considered. That is, to try to analyze and design the controller without using a precise mathematical model of the plant; to try to obtain results which are not only valid for the nominal model but also for a given range of models of the plant, and consequently for the real process.

When designing controllers using linear and time invariant models of the plant, there are two possible approaches for contemplating the uncertainties in the system model and the disturbances around it [52]. The first of these is to use an adaptive controller, which estimates the parameters and calculates the control signal for those parameters.

The second approach is to consider the system uncertainties in the design of a fixed controller, which leads to a robust control scheme less sensitive to variations in the parameters and the perturbations. There are various strategies for carrying out this scheme. One of them is that of calculating a linear quadratic gaussian controller LQG with transfer function recovery of the linear quadratic regulator LQR, a scheme known as LQG/LTR. Another possible way is to minimize the worst possible case which is measured by a certain norm of a transfer function (usually H_∞ norm).

There are other possibilities which combine the design taking into account the model uncertainties with adaptive control, as is analyzed in chapter 5.

1.6.1 The robust control problem

Real processes are too complex to be described in an absolute and precise manner by a mathematical model. There are always discrepancies between processes and their models (modelling errors). Added to this is the fact that the aim is to describe the system with a linear time invariant model implying a set of simplifying hypothesis which augment the errors of the original model. Among the main factors causing modelling errors are:

1. Modifications of the working point of the plant.
2. Unaccounted nonlinear dynamics.
3. Unmodelled high frequency dynamics.
4. Unaccounted delays.
5. Imprecisions in the parameters, due to the identification method and/or the model used.

Taking into account how these factors affect the model errors, they can be classified into two large groups: the *parametric* uncertainties (1 and 5) and the *structural* ones (2, 3 and 4). With regards to the available information about the causes of uncertainties these can be divided into structured and unstructured uncertainties.

In the case of structured uncertainty, it is only known that there are discrepancies between the model and the real plant, and it may also be possible to know the maximum size of the deviation of determined input/output measures (for example, the biggest discrepancy in the frequency response caused by unmodelled high frequency dynamics or the difference in the time response).

In the case of structured uncertainties, it is known that the uncertainty is to some extent due to certain differentiated elements of the plant, in the form of tolerances of their values (for example, the uncertainties in the value of a pole and/or a zero). It is also possible to have a partial knowledge from the uncertainty sources in which case it is possible to talk about partially structured uncertainties.

The objectives of control are that the designed controller should work well when implemented in the real process in spite of modelling uncertainties. It is obvious that the system should be stable in closed loop for the nominal model. This is known as *Nominal Stability* (NS). On the other hand, once stability has been achieved it is necessary for certain system variables to present adequate and in some cases optimal behaviour in respect to a cost function or behaviour index. In this case *Nominal Performance* (NP) is aimed at.

A further requirement looked for in control systems is that they should be stable in closed loop for the set of possible plants which can arise as a consequence of the uncertainty in the plant model. This objective is known as *Robust Stability* (RS). In some cases it is not enough for the control system to remain stable but certain performance specifications have to be fulfilled for all the possible plants (*Robust Performance* RP).

The control of systems with uncertainties enters within the discipline known as *Robust Control.* This discipline was mainly developed in the eighties. The most outstanding methods are H_∞ methods by Doyle *et al.* [41], LTR methods by Doyle and Stein [43] and Stein and Athans [114], Kharitonov methods by Barmish [8] and μ synthesis methods by Balas *et al.* [7].

1.6.2 Control methods in the frequency domain

Frequency control schemes are based on the use of graphic methods and on the representation of the system performance in the frequency domain. These methodologies were developed during the thirties and forties and yet they still apply today. Acceptable results can be obtained with them for the four problems described at the beginning of the section (stability, noise immunity, specifications of the closed loop transitory performance and improvement of robustness to variations in the plant parameters).

However they have two types of deficiencies. On one hand, they only work well for single input single output (SISO) systems (although currently great efforts are being made to extend their use to the study of multivariable systems [104], as in the case of H_∞ type of methodologies); on the other hand, they are not analytical methods and thus they do not lead to implementations of an algorithmic type.

Amongst the basic advantages of frequency control methods the following stand out:

a) Performance specifications can be directly translated to the frequency domain.
b) In most cases it is easer to obtain a frequency response model than a parametric one.
c) Modelling errors and uncertainties can be easily and directly interpreted in the frequency domain.
d) Because frequency design methods are of a graphic and approximate nature they are inherently robust.

This capacity to consider robustness problems is one of the reasons for the renewed interest in the study of frequency domain methods in the literature ([43], [106], [66], etc.).

New ideas have recently appeared, aiming to relate adaptive control to robust design. This link is established in the frequency domain by considering an adaptive control identifier as a mechanism which reduces the level of uncertainty about the plant being controlled ([68], [62], [50]). Design methods in the frequency domain used in adaptive control have also appeared recently in the literature. In [53] a relay in the feedback loop is used to obtain the gain and frequency at the point of the Nyquist curve where the process has a phase lag of π radians. This information is then used to adjust a PID controller. In [83] a frequency approximation for systems which can be described by the step truncated response is used and various controller design techniques are suggested. In [118] adaptive control algorithms are presented adjusted to fulfill specifications in the frequency domain. In [57] an adaptive control scheme is presented combining frequency domain estimation with controller design also in the frequency domain.

All these algorithms use an unitary feedback configuration and have the property that the prior knowledge required about the plant is very small and that the plant frequency response is estimated directly, allowing for the implementation of robust adaptive controllers.

In this book, an Internal Model Control (IMC) structure is used instead of the classical unitary feedback configuration to develop a frequency based adaptive IMC structure. The IMC structure corresponds to a Youla-Kucera [3] parameterization of all stabilizing controllers and has gained wide popularity after the book of Morari and Zafiriou [86] showed the benefits of using IMC. Many applications of IMC to diverse types of processes have appeared since (see [49], [51] and [11]).

A simplified diagram of the IMC control structure can be seen in Figure 1.6. The process is represented by P, while P_m is a model of the process and Q represents the controller. If there are no modelling errors ($P_m = P$), and there are no external perturbations ($d = 0$), the output (y) of the plant coincides with the output of the model and as there is no feedback signal, then:

1. The controller (Q) can be designed in a open loop manner and
2. The resulting control structure is stable if, and only if, the process is open loop stable and the controller is also stable. The feedback signal is included to account for uncertainties and disturbances.

1.6.3 Optimal control methods

Modern control methods have an essentially algebraic character so that they benefit from algorithmic computer implementations, allowing for design aid packages to be used. They can also be applied to systems of certain complexity, both multidimensional and nonlinear. Because of these advantages it would seem that they should have been able to displace the classical methods. This is not so, because although modern methods allow for the first three problems (stability, noise immunity and transitory specifications) to be resolved reasonably well, up to date, the solutions found for the fourth problem

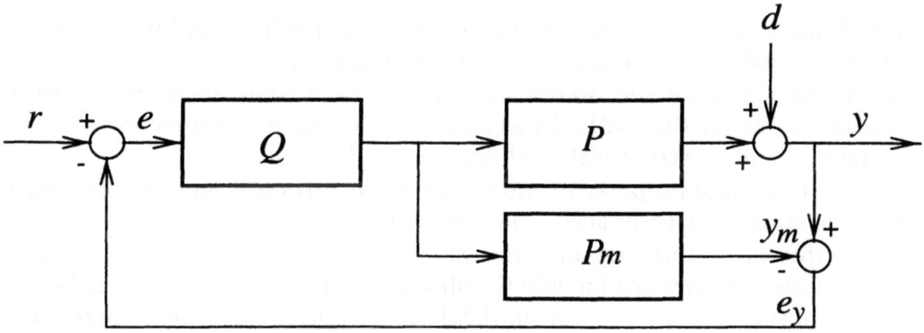

Fig. 1.6. Internal model control structure (IMC)

(robustness) are not as good as those provided by the classical methods. The stability margins (*phase* and *gain margins*) play an important role in these latter methods. These margins allow for a very simple geometrical interpretation using the graphic representation of the transfer function in the frequency domain and they measure the distance of the system from the point at which stability will be lost. Essentially, these are indexes that measure the relative stability of the system and they were introduced as such originally. However, recently it has been seen that they also constitute a notable robustness measurement. If they attain reasonable values then the performance varies only slightly when there are changes in the parameters (aging and bad estimation of components, etc.) or in the system structure (wrong models, saturations, etc.). The farther the system from the critical stability point is the greater its capacity to support variations of the parameters with no noticeable effect on the performance; that is, the system will be more robust against parameter variations.

The stability margin concept is fundamental in classical methods and up to date no index playing an analogous role has been found in modern methods of control system design. This, amongst other things, explains the survival of classical methods.

Modern methods of control system design were developed during the fifties and sixties. The most elaborate result of these methods is the LQG. This method is based on conventional hypothesis: the system is linear, the criterion quadratic and the random disturbances are gaussian. The computation of the regulator is done completely automatically by a computer. This method is based on an internal description of the system, on the state concept (as necessary and sufficient information for making decisions in relation to the future evolution of a system), and on the separation principle of estimation and control. All these elements provide an extremely powerful method for solving the design problem of a control system. A separated solution to the problem is given: firstly a state estimation is obtained by an observer in the deterministic case or by a *Kalman filter* in the stochastic one. A linear control law is applied to this estimation which is fed back and with this the complete regulator is obtained. It has traditionally been noticed that the LQG method worked very well when very precise mathematical models were available. However the method was extremely sensitive to imprecisions in the parameters and to structural modifications.

An outstanding property is that the closed loop transfer function obtained when the state is directly accessible (thus without observer or *Kalman filter*) is the same as if this last module was included. That is, the closed loop transfer function does not include the observer. This fact led to the conclusion that the observer was irrelevant in the design of the closed loop system. This erroneous belief lasted for more than a decade. The truth, discovered at the end of the seventies, is that although it is certain that the observer has no effect on the transfer function in closed loop it has a strong influence on the open loop transfer function. Furthermore, this influence is enormously harmful to the system robustness (in fact, as will be analyzed in chapter 6, it can make the stability margins noticeably decrease).

The relevance of the discovery that the introduction of the observer deteriorates the system robustness is easier to understand if one considers the well known and remarkable result of the LQR regulator, thanks to which if the state (without observer) is directly fed back, the system will have a phase margin of 60 degrees and a gain margin which asymptotically tends to infinity.

In order to solve the problems presented by the deterioration of robustness caused by the introduction of the observer the method known as LQG/LTR has been proposed amongst others. This method aims at using the analytical possibilities of LQG (and in consequence the use of the computer in aided-design) but modifying the *Kalman filter* (observer), so that the harmful effects on the robustness (on the stability margins) are attenuated, making the open loop transfer function of the system with observer asymptotically approximate the one which this would be if the observer was not included. For this reason the controller developed in the chapter devoted to frequency domain and robust controllers has included the LTR mechanism, given that the recovery mechanism of the transfer function in open loop aims at recuperating eminent characteristics related to the frequency domain methods, such as the stability margins.

1.7 Artificial Intelligence Techniques

As pointed out by Antsaklis [113], when designing controllers for complex dynamical systems, it should be desirable to incorporate functions to the control system which have been performed in the past by human operators. Conventional control theory may address many control problems, but often heuristic methods are necessary for controlling complex processes. This type of control is referred to in literature as intelligent control.

Amongst this type of techniques, knowledge based systems (KBS) also called *"expert systems"*, neural networks (NN) and fuzzy logic controllers (FLCs) can be mentioned.

The third strategy (FLC) is treated in the book in order to show a nonlinear heuristic control structure based on the authors' experience of operation with the plant. Strictly speaking, the technique cannot be considered a "pure" intelligent control strategy (as intelligent controllers are designed to emulate mental faculties such as adaptation and learning, and to cope with uncertainty, large amounts of data etc.) but an approximation in which the experience in the operation with the plant is translated into a fuzzy logic control framework. In this way, knowledge about the process can result in a control structure.

Fuzzy logic provides a conceptual base for practical problems where the process variables are represented as linguistic variables which can only present a certain limited number of possible values (for example, *very large, large, normal, small,* etc.). The linguistic variables can then be processed using a series of rules.

There is a certain amount of controversy in the field of fuzzy control about the possibility of its application (such as those mentioned in the article by Aström [4]), about when and how it should be applied and about the impossibility of reconstructing the development process of this type of controller. Another problem is that of the inexistence of a theoretic basis justifying the properties of closed loop with this type of controller.

Fuzzy logic control seems to be appropriate when working with a certain level of imprecision, uncertainty and partial knowledge, and also in cases where the knowledge of operating with the process can be translated into a control strategy that improves the results reached by other classical strategies that do not take into account this knowledge about the process. In this book fuzzy control has basically been used to develop nonlinear controllers which permit the control signal to be adapted to the changes in the process dynamics due to changes at the operating point or to the disturbances.

CHAPTER 2
DESCRIPTION AND DYNAMIC MODELS OF THE PLANT

The objective of this chapter is to describe the distributed collector field and the models used for its simulation and control. The process considered is the distributed solar collector field ACUREX of the "Plataforma Solar de Almería" (PSA). This is an installation which belongs to the Spanish "Centro de Investigaciones Energéticas, Medioambientales y Tecnológicas" (CIEMAT) of the Ministry of Industry, Commerce and Tourism, and is jointly operated by the CIEMAT and the "Deutsche Forschungsanstal für Luft-und Raumfahrt" (DLR), of Germany. It is the largest European test center for solar thermal energy applications.

There exists an increasing interest in the use of the so called *renewable* or *alternative energies*. Such is the case of solar energy, existing many solar plants all over the world. Particularly, the thermosolar electrical plants (TEPs) are systems employed to obtain electrical energy from solar energy by the previous transformation of this into thermal energy. The design of this kind of plant started in 1977 and the first operating plant started operation in 1981 [105].

The main difference between a conventional plant and a solar plant is that the primary energy source, while being variable, cannot be manipulated. Due to this fact, a solar plant is required to cope with some problems which are not encountered in other power plants.

Two main types of TEPs can be mentioned, depending on the number of components which transform the solar energy into thermal energy: if there is only one component, the systems are called central receiver systems (CRS). If each collector incorporates its own transformation device, this configuration is called a distributed collector system (DCS). The text focuses on the DCS type of plants.

2.1 Plant description

The solar power plant SSPS (Figure 2.1) is located at the Plataforma Solar de Almería. The plant was originally designed for the production of electricity with a CRS and a DCS composed of two distributed collector fields:

- The ACUREX field, with elevation solar tracking system and Acurex collectors model 3001.
- The MAN field, with elevation and azimuth solar tracking system.

Fig. 2.1. Solar power plant SSPS

Figure 2.2 shows a diagram of the power generation system in which both fields are represented. The plant used to implement the advanced control strategies was the ACUREX field, which is nowadays used as a test-bed plant by different European companies and universities. During the last six years, advanced control structures have been tested on this field, many of which are included in this book.

Figure 2.3 shows a photograph of the ACUREX field. The distributed solar collector field consists of a series of mirrors that reflect solar radiation onto a pipe where oil gets heated while circulating. It can provide 1.2 MW_t of peak thermal power with a beam solar radiation of 900 W/m^2. The daily thermal energy delivered by the field is about 6.5 MWh_t [127]. The field consists of 480 distributed solar collectors. These collectors are arranged in 20 rows which form 10 parallel loops (each of the loops is 172 metres long).

The collector uses a parabolic surface to concentrate a direct beam of solar radiation onto the receiver tube, which is located at the focal point of the parabola. The heat transfer fluid is pumped through the receiver tube and picks up the heat transferred through the receiver tube walls. These tubes use the concentrated solar radiation to heat thermal oil, which is extracted from the bottom of a thermal storage tank, circulated through the collector field and returned to the top of the tank using a pump located at the field inlet. Due to the good thermal stratification properties of the storage tank and to the *thermocline effect* the oil at the top of the tank can be maintained at a high temperature during a few days. The system is provided with a three way valve which allows the oil to be recycled in the field until its outlet temperature is adequate to enter into the storage tank. A simplified diagram of the collector field can be seen in Figure 2.4.

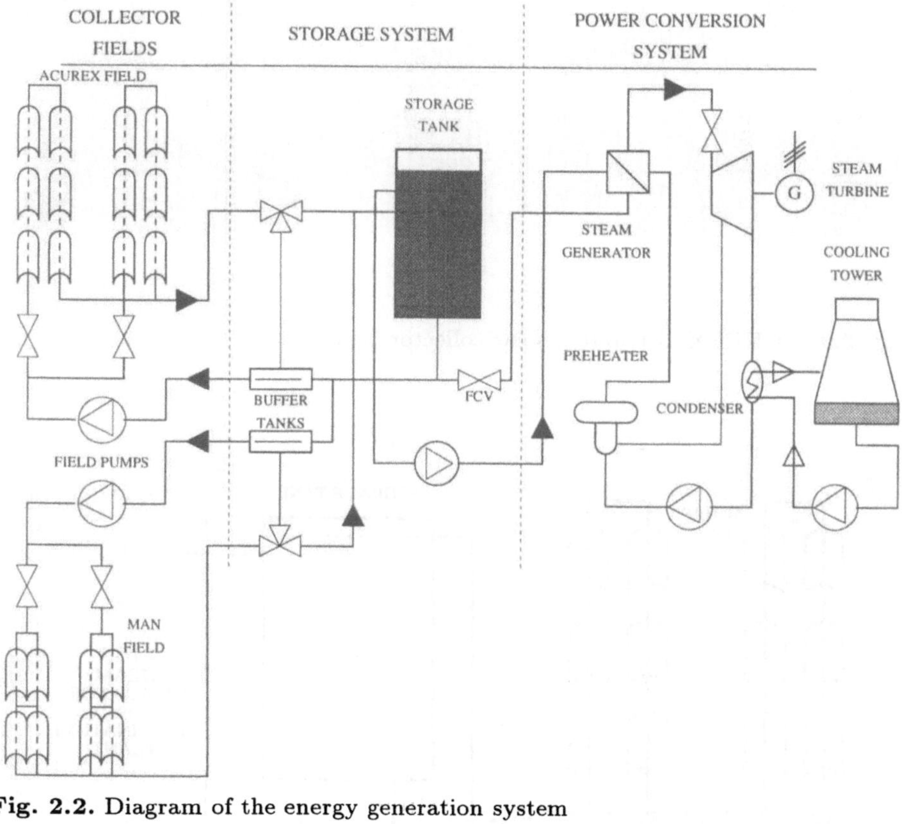

Fig. 2.2. Diagram of the energy generation system

The storage tank is connected to various systems. Particularly, the tank is connected to a steam generator so that, making use of the thermal stratification principle, hot oil is extracted from the top of the tank to produce steam and cold oil is returned to the bottom of the tank. The steam is used in a 7-stage turbine connected to the electrical generator to produce electricity. The hot oil can also be used to feed the low pressure evaporator of

Fig. 2.3. ACUREX distributed solar collector field

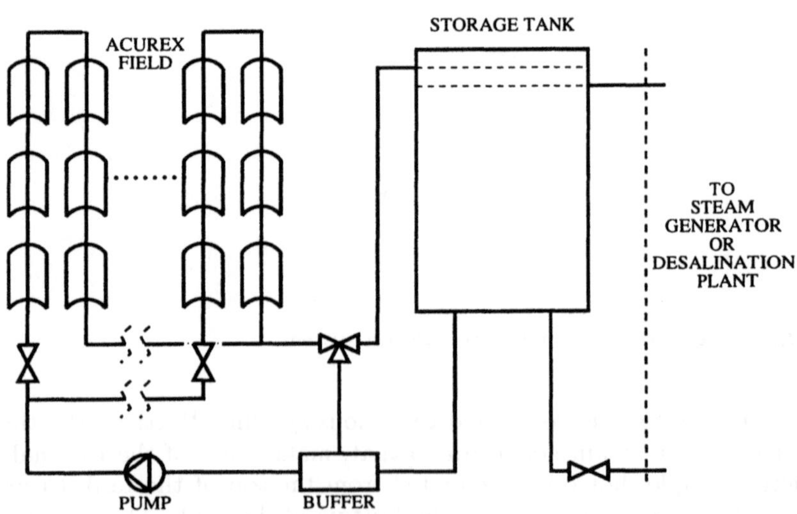

Fig. 2.4. Schematic diagram of the collector field

a desalination plant to produce drinkable water from seawater in a 14-stage destilation process. The hot oil produces steam at 70°C and 0.35 bar, which condenses inside the first evaporator of the desalination plant, delivering the energy needed by the desalination process [127].

The field is also provided with a sun tracking system which causes the mirrors to revolve around an axis parallel to that of the pipe. The tracking system is very simple and is based on the voltage differential which appears between two photodiodes or photovoltaic cells placed on the collector axis in a plane perpendicular to the sun rays. A plaque is placed between the photodiodes in such a way that if the plane of the cells is inclined in respect to the perpendicular of the sunrays, a shadow is projected onto one of the cells, generating the differential voltage which, adequately amplified, produces the motors to move the collectors to the adequate position. This simple sun tracking system, schematically represented in Figure 2.5 has been treated in some basic classical control books [93].

The seeking mechanism can reach three possible states:

Track: The mechanism seeks the sun and the collectors focus on the pipe.

Desteer: The mechanism steers the collector several degrees away from the sun and continues tracking with the receiver out of focus. This protects the field from over-heating in case of a pump failure.

Stow: The mechanism moves the collector to an inverted position at the end of each day or if a serious alarm occurs.

The system takes about 5 minutes to take the field from stow to track and only 20 seconds from desteer to track. A more detailed description of the field can be found in [59].

2.2 Objective of the control system

The objective of the control system in a distributed collector field is to maintain the outlet oil temperature of the loop (or the highest outlet oil temperature reached by one of the collectors each sampling time) at a desired level in spite of disturbances such as changes in the solar irradiance level (caused by clouds), mirror reflectivity or inlet oil temperature.

A fundamental feature of a solar power plant is that the primary energy source, solar radiation, cannot be manipulated. The intensity of the solar radiation, in addition to its seasonal and daily cyclical variations, also depends on atmospheric conditions such as cloud cover, humidity, and air transparency. It is important to be able to maintain a constant outlet temperature for the fluid as the solar conditions change, and the means available for achieving this is via the adjustment of the fluid flow.

These daily solar power cycle characteristics are such that the oil flow has to change substantially during operation. This leads to significant variations in the dynamic characteristics of the field, such as the response rate and the

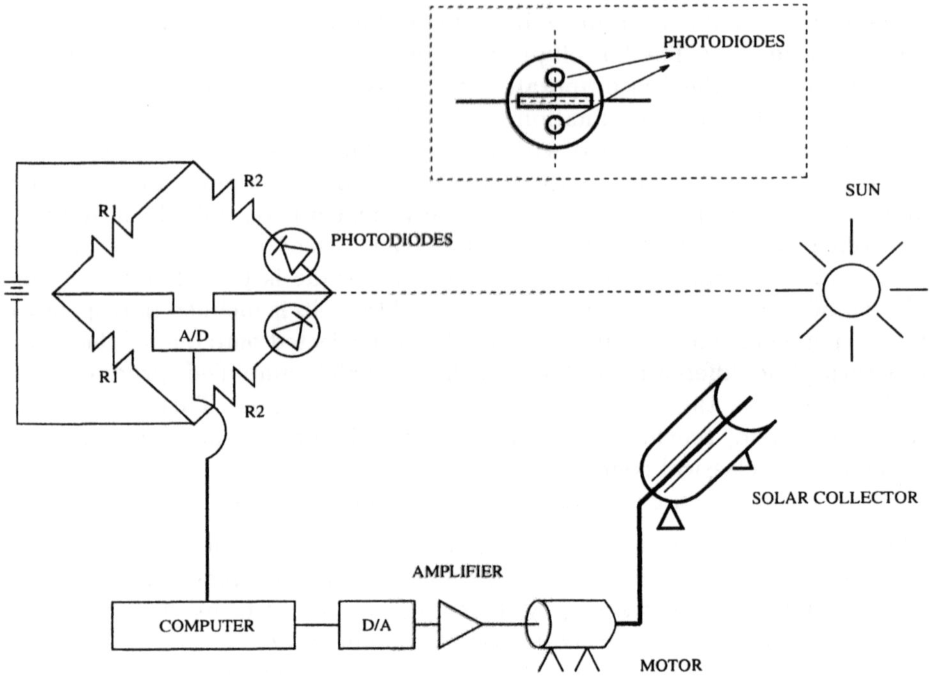

Fig. 2.5. Working principle of the tracking mechanism

dead time, which cause difficulties in obtaining adequate performance over the operating range with a fixed parameter controller.

2.3 Data acquisition system

The available equipment for control purposes is constituted of a 486/66MHz PC with two screens. In one of these screens, those variables which are meaningful from the power generation viewpoint are presented in real time (outlet oil temperatures of each one of the loops, pressure in the heat exchangers inlet, etc.). The other screen is reserved to display the evolution of the fundamental variables of the control algorithms used at the plant. This is a common configuration in this type of plant. A data acquisition system runs in the PC. Each 3 seconds more that 150 data are registered and processed. The minimum sampling time allowable is 3 seconds, which is a small value compared to the appropriate sampling time for control purposes (39 seconds).

The system is also provided with a supervisory module which manages the different alarms which can be produced during the daily operation (this module sends mirrors to the stow position if in any of the loops the outlet temperature exceeds 300°C, if there are oil losses, etc.).

The direct control system of the pump which impels the fluid is an analog PI adjusted to the design point corresponding to 12 hours (solar hour), March the 21st with direct beam radiation of 920 W/m^2. As can be understood, the environmental conditions under which a solar plant has to operate are very different from the design conditions.

Due to the nature of solar energy, its availability depends on the date and hour, environmental conditions and other disturbances. This justifies the necessity of developing advanced control systems which allow the field to maintain a stable outlet oil temperature under a wide range of operating conditions.

As a complement of the data acquisition system and the analog PI, diverse control schema have been developed so that the distributed solar collector field can be operated without human control and supervision. The control program [13] is embedded in the data acquisition system.

2.4 Dynamic simulation models of the field

The development of numerical simulation models of the plant has played an important role in the design of different control strategies avoiding a number of expensive and time consuming controller tuning tests at the solar power plant. The availability of a model which describes the main dynamics of the field allows the designer to simulate a great diversity of operating conditions, disturbances, etc., leading to pre-designs which can produce acceptable results when implemented at the actual plant. Small modifications to obtain good control system performance are needed when commissioning the controllers. As the starting phase of the operation takes a long time, if the controller produces non-acceptable results, significant part of the daily operation can be lost.

Two simulation models have been developed and programmed: a concentrated parameter model and a distributed parameter model. These models are based on the work done by Carmona [29].

2.4.1 Concentrated parameter model

Supposing a concentrated description of the plant, the variation in the internal energy of the field can be described by:

$$C\frac{dT}{dt} = n_o SI - \dot{q}Pc_p(T - T_{in}) - H_l(T_m - T_a) \qquad (2.1)$$

where T is the outlet temperature of the model in a determinated time instant, I is the effective solar radiation, T_{in} is the inlet oil temperature of the field, T_m is the mean inlet-outlet temperature, T_a is the ambient temperature, $\dot{q}$ is the oil flow, n_o is the collector optical efficiency, S is the effective surface, C is the specific thermal capacity of the fluid, Pc_p is a term that accounts for the product and quotient of characteristic magnitudes (areas, thermal capacities, etc.) and H_l is the global thermal losses coefficient.

The product Pc_p (function of the temperature) can be approximated by 1924 KJ/l°C. By using experimental data, the global thermal losses coefficient H_l has been obtained to be 1.05 KW/°C. These values were obtained by applying the multiple regression method to a set of 60 measurements obtained under different operating conditions in steady state [29].

In order to identify the other parameters which appear in the concentrated formulation a square wave signal was introduced in the oil flow and input-output data generated were used to perform the adjustment of the parameters using the least squares method. As a result of the identification, the parameters were obtained to be $C = 2267$ KW/°C and $n_o S = 1322$ m^2. The results of the adjustment of this model can be found in [29] and [99].

2.4.2 Distributed parameter model

As has been mentioned, each of the ACUREX loops is formed by four twelve module collectors suitably connected in series. The loop is 172 metres long, the active part of the loop (the parts where the solar radiation reaches the tube) measuring 142 metres and the passive part (joints and other parts not reached by concentrated solar radiation) 30 metres. Due to the complexity of the system and the existence of nonlinearities, a numerical model has been evolved to simulate it.

The dynamic behaviour of the distributed collector field is simulated by 100 lumped parameter submodels. Temperatures of the oil and the tube walls are modelled separately. The model also takes into account sun position, field geometry, mirror reflectivity, solar radiation and the inlet field oil temperature [25], [14].

The collector loop is the basic subsystem which determines the behaviour of a collector field. If it is possible to model a loop, the behaviour of the whole field can be determined by simply adding the parallel loops and allowing for transport delays in the interconnecting tubes. The present model has been developed to fulfill the following objectives:

- The simulation of the field behaviour in order to optimize the temperature regulation system.
- A study of the behaviour of the system under specific operational conditions such as in cases of passing clouds.
- A study of extreme situations by simulation of failures, desteer mechanisms, etc.
- The application to other collector fields by modifying the corresponding parameters.

The model allows to know the temperature distribution in the absorption tube and in the thermal oil along the collector loop at a given moment, as well as the temporary variation of the temperatures at determined points of the collector. The following hypotheses have been made:

- The properties of the oil are considered as functions of the temperature.
- The flow in each section is presumed to be circumferentially uniform and equal to the average value.
- Variations in the radial temperature of the tube walling are not taken into account. This assumption is reasonable in the case of a thin wall with good thermal conductivity.
- The oil flow and the irradiance are considered as time functions and are always the same for each element (an incompressible fluid is presumed).
- Losses caused by the conduction of axial heat on both sides of the wall and from the fluid are negligible. The axial conduction in the tube should be slight given that the walling is thin, having a high heat resistance. The axial conduction in the fluid is relatively slight as the oil conductivity is poor.

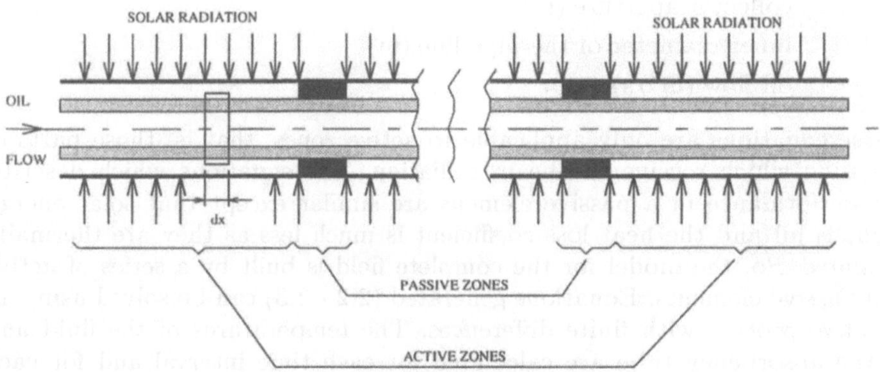

Fig. 2.6. Element of the distributed parameter model

Using the above hypotheses and applying the conservation of the energy in the metal tube of a length control volume dx (Figure 2.6) over a time interval dt:

$$\rho_m C_m A_m \frac{\partial T_m}{\partial t} = I n_o G - H_l G (T_m - T_a) - L H_t (T_m - T_f) \tag{2.2}$$

and for a fluid element:

$$\rho_f C_f A_f \frac{\partial T_f}{\partial t} + \rho_f C_f \dot{q} \frac{\partial T_f}{\partial x} = L H_t (T_m - T_f) \tag{2.3}$$

In the above equations the subindex m refers to the metal and that of f to the fluid and:

ρ : oil density (Kg/m^3)
C : field capacity (KJ/Kg^oC)
A : transversal area (m^2)
T : temperature (^oC)
T_a : ambient temperature (^oC)
I : irradiance (W/m^2)
n_o : optical efficiency
H_l : global coefficient of thermal losses ($W/m^{2o}C$)
This coefficient is considered per unit of mirror area, that is, the power in watts lost per mirror square meter and Celsius degree
H_t : coefficient of metal-fluid transmission ($W/m^{2o}C$)
This corresponds to the thermal power lost through the contact surface between fluid and metal
G : collector aperture (m)
L : inner diameter of the pipe line (m)
$\dot{q}$: oil flow (m^3/s)

These equations are only applicable to active zones, that is, those parts of the tube which receive the beam radiation. The equations which describe the performance in a passive element are similar except that solar energy input is nil and the heat loss coefficient is much less as they are thermally insulated. So, the model for the complete field is built by a series of active and passive elements. Equations generated (2.2 - 2.3) can be solved using an iterative process with finite differences. The temperatures of the fluid and of the absorbency tube are calculated for each time interval and for each element. Each segment is 1 metre long and the integration interval is 0.5 s. A two stage algorithm has been chosen to solve the temperature equations. In the first stage the temperatures of the fluid and of the metal are calculated supposing that the fluid is in a steady state. In the second stage the fluid temperature is corrected in function of the net energy transported by the fluid.

– 1st Stage

$$\begin{aligned} T_m(n,k) &= T_m(n,k-1) + \frac{\Delta t}{\rho_m C_m A_m}(In_oG - H_lG(T_m(n,k-1) - T_a) \\ &\quad -LH_t(T_m(n,k-1) - T_{1f}(n,k-1)) \end{aligned}$$

$$T_f(n,k) = T_{1f}(n,k-1) + \frac{LH_t\Delta t}{\rho_f C_f A_f}(T_m(n,k-1) - T_{1f}(n,k-1))$$

– 2nd Stage

$$T_{1f}(n,k) = T_f(n,k) - \frac{\dot{q}\Delta t}{A_f\Delta x}(T_f(n,k) - T_f(n-1,k))$$

In these difference equations, $T_f(n,k)$ and $T_m(n,k)$ are the temperatures of fluid and metal in segment n during the k_{th} time interval.

The different constants and coefficients used in the previous equations have been determined using real data from the plant. Many of them being adjusted to polynomial functions of the temperature by a least squares method.

Properties of thermal fluid. The fluid used to transport the thermal energy is Santotherm 55, a thermal oil which allows working temperatures above 300°C without decomposing. One of the main characteristics of this oil is its low thermal conductivity. Furthermore, its density is highly dependent on its temperature, which permits the use of just one storage tank to contain both the hot and the cold oil in thermal stratification (the thermocline effect). The tank used in this collector field has a capacity of 140 m^3 which allows for the storage of 2.3 thermal MWh for an inlet field temperature of 210°C and an outlet field one of 290°C. The good thermal stratification of this oil allows it to be stored for various days.

From data supplied by the oil producer [59] its physical properties have been obtained as the following polynomial function of the temperature T (°C):

a) Density: $\rho_f = 903 - 0.672\,T$ (Kg/m^3).
b) Specific thermal capacity: $C_f = 1820 + 3.478\,T$ (J/Kg °C).
c) Thermal conductivity: $K_f = 0.1923 - 1.3\;10^{-4}\,T$ (W/m°C).
d) Dynamic viscosity: $\mu = 1.41\;10^{-2} - 1.6\;10^{-4}\;T + 6.41\;10^{-7}\;T^2 - 8.66\;10^{-10}\,T^3$ (Pa.s).
e) Prandlt number: $Pr = 212 - 2.2786\,T + 8.97\;10^{-3}\,T^2 - 1.2\;10^{-5}\,T^3$.

Thermal Transference Coefficient. This coefficient can be expressed as a function of two variables: the flow ($\dot{q}$) and a coefficient depending on the temperature (H_v).

$$H_t = H_v\dot{q}^{0.8}$$

$H_v(T)$ has been evaluated for different temperatures and by adjusting the values obtained to a polynomial, one gets that:

$$H_v(T) = 2.17\;10^6 - 5.01\;10^4\;T + 4.53\;10^2\;T^2 - 1.64\;T^3 + 2.1\;10^{-3}\;T^4$$

Global Coefficient of Thermal Losses. The losses have been evaluated by various tests carried out on the ACUREX field with the oil circulating. In the steady state the losses are calculated by multiplying the enthalpy lost in the oil by the masic flow. Using the data, it has been established that the total amount of the losses can be approximated by:

$$L_t = 0.00667\,\Delta T_m^2 - 0.164\,\Delta T_m\ \text{(KW)}$$

where ΔT_m is the difference in temperature between the average inlet and outlet temperature and the atmosphere temperature. The loss coefficient is given by:

$$H_1 = L_t/A\,\Delta T_m \quad ; \; A = 2674\ \text{m}^2\ \ \text{(total collector surface)}$$

that is,

$$H_l = 0.00249\,\Delta T_m - 0.06133\ \ (\text{W/m}^{2\,o}\text{C})$$

In the model the thermal loss coefficient is calculated by applying the above equation to each element of length and having ΔT_m equal to the temperature of this element minus the ambient temperature.

Determination of the effective surface. The amount of shadows produced by each mirror on the others and therefore the effective surface of the collector field depends on the position of the sun. To take into account these effects the position of the sun is first calculated based on the solar hour, Julianne day (JD), the year and the local latitude. The incidence angle ϕ (angle between the solar vector and the vector perpendicular to the collector surface) is given by:

$$cos\phi = (1 - cos^2\delta_l sin^2\delta_2)^{1/2}$$

with $\delta_2 = 15(T_s - 12)$ being the hourly angle; T_s the solar hour and δ_l the declination calculated by:

$$\delta_l = 23.45 sin(360(284 + JD)/365)$$

The effective surface is given by: $S = S_{ref}E_o$, $E_o = 1-(S_1+S_2 tg\phi)/S_3$ where E_0 is the global optical efficiency (with coefficients S_1, S_2 and S_3 related to the collector dimensions) and S_{ref} (m^2) is the reflecting surface of the mirrors of the ACUREX field. A more detailed description of the model can be found in [29] and [14].

2.4.3 Model validation

A study in which results obtained with the simulation model are compared to real experimental data can be found in [25]. In these experiments the response of the simulator is compared to that of the real plant under many different conditions (changes from track to desteer position, from desteer to track position, response to oil flow steps, etc.).

As an illustrative example, Figure 2.7 shows the output of the model in an open loop configuration and that obtained with the real system under the same conditions of oil flow, solar radiation and inlet oil temperature (without any intermediate correction using the real plant output). As can be seen, after an initial transient in which temperatures of the model and real system differ substantially (mainly due to the fact that it is impossible to know the initial temperature profile both in the tubes and at the lower thermocline of the storage tank and to the degradation of the thermal properties of the oil at low temperatures), the model tends to follow the system behaviour, with a logical difference which tends to zero as the system approaches the steady state. At the end of the operation (which normally finishes at four in the afternoon - local hour), a discrepancy appears between the outlet of the model and that of the plant due to the progressive decreasing of the solar radiation, because the coefficients of the model were obtained for more operational (higher) solar radiation.

The temperature increase experimented by both the model and the field at about 14.25 decimal hours was due to the inclusion of a loop that was not operative at the beginning of the operation. Due to this fact, the oil flow which was circulating through each loop experimented a sudden decrease. The second significant change in the oil flow signal and indeed in the output of the system was produced at about 15.4 decimal hour, due to a set point change. As can be seen, the behaviour of the nonlinear distributed parameter model in both situations is very similar to that of the plant.

From the control viewpoint, in order to assure a secure operation, the controlled variable is the outlet temperature of the loop with the highest value at each sampling time, and not the global outlet oil temperature. This fact can introduce additional dynamics if the highest temperature loop changes occur during the operation, but avoids any of the loops exceeding the maximum allowable temperature, in which case the mirrors are sent to stow position by the supervisory system, causing heavy losses in the daily operation.

2.5 Analysis of the dynamic response of the plant

In order to analyze the dynamic behaviour of the distributed solar collector field, step response and pseudo random binary sequence tests (PRBS) [74] have been performed to obtain the dynamical information needed for modelling purposes.

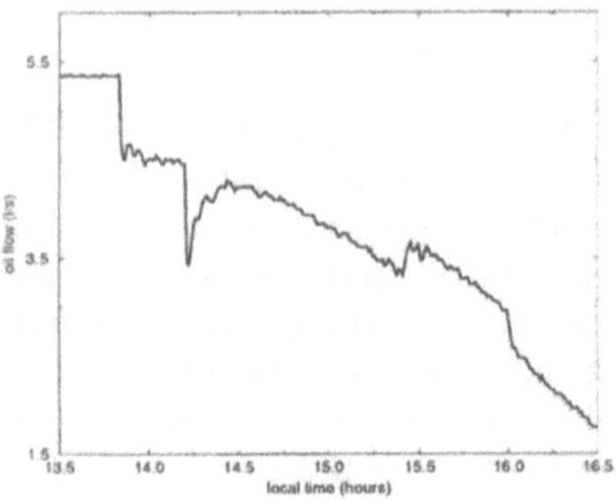

oil flow (l/s)

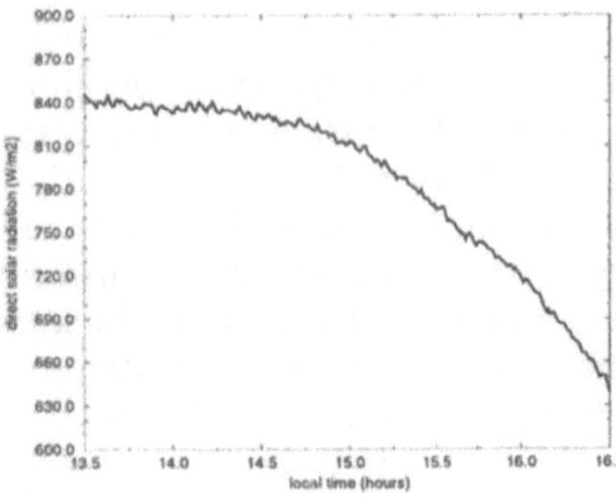

direct solar radiation (W/m^2)

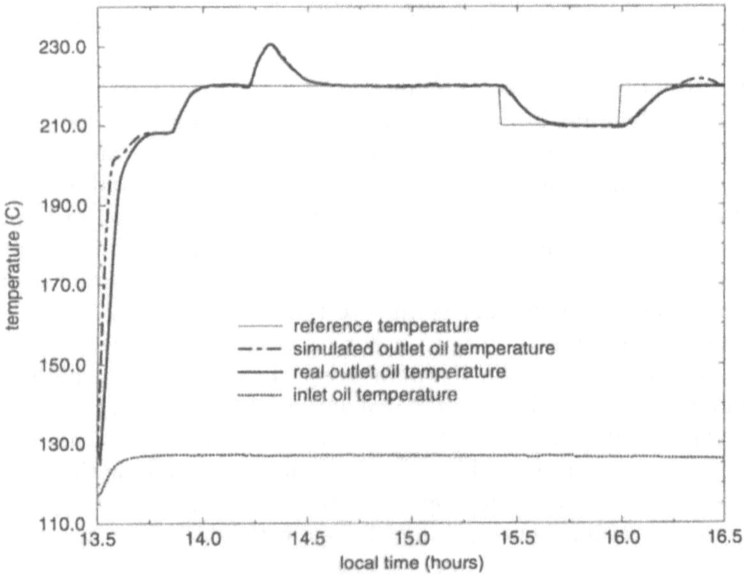

set point, inlet and outlet oil temperatures (°C)

Fig. 2.7. Comparison simulator-plant

2.5.1 Analysis of the time response

One of the tests carried out to characterize the field consists of introducing a step input signal in the oil flow in an open loop configuration. The response to a change in oil flow from 8 l/s to 7 l/s can be seen in Figure 2.8 (at noon). As can be appreciated, the response can be approximated by that of a first order system or overdamped second order system with a delay. This kind of step response suggests the use of low order linear descriptions of the plant (as is usually done in the process industry) to model the system and to design diverse control strategies.

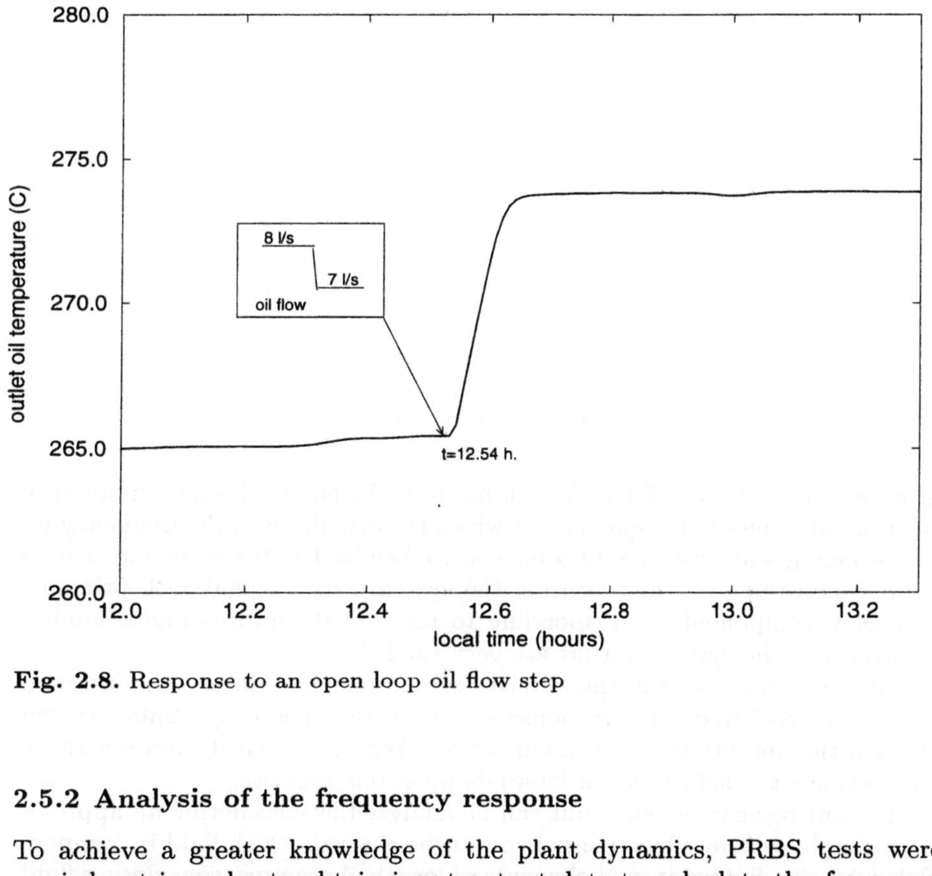

Fig. 2.8. Response to an open loop oil flow step

2.5.2 Analysis of the frequency response

To achieve a greater knowledge of the plant dynamics, PRBS tests were carried out in order to obtain input-output data to calculate the frequency response of the plant corresponding to different operating conditions.

Figure 2.9 shows the frequency response[1] obtained by a spectral analysis of the input-output data for a PRBS test on the plant corresponding to a

[1] The frequency response corresponds to the system constituted by the feedforward studied in chapter 3 in series with the plant.

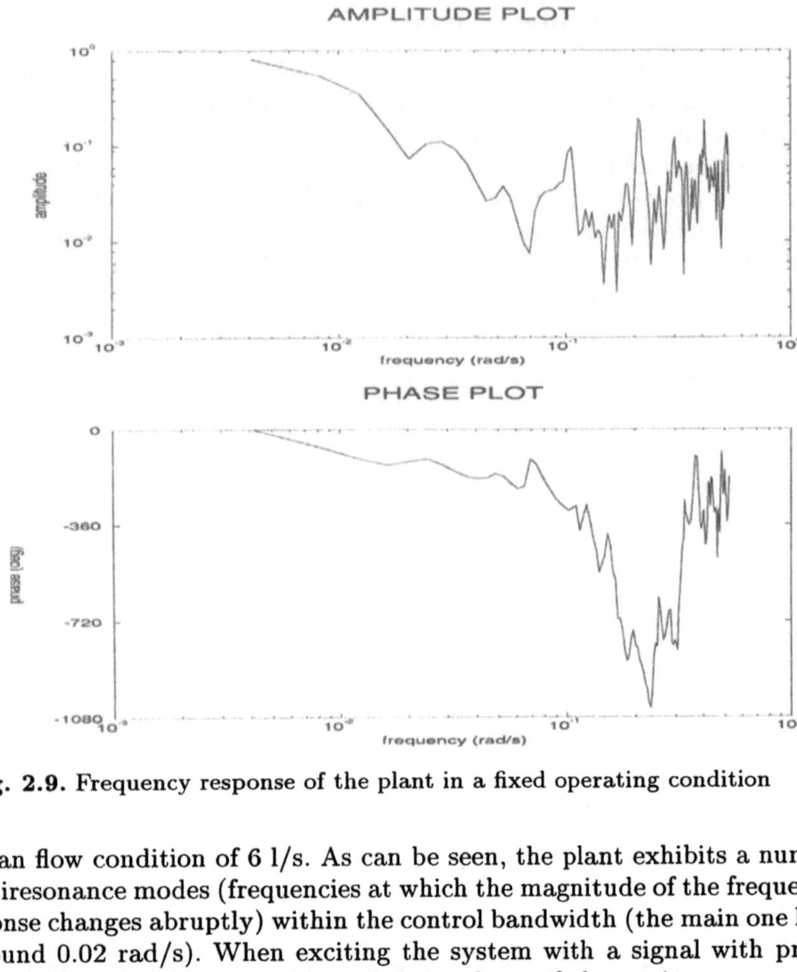

Fig. 2.9. Frequency response of the plant in a fixed operating condition

mean flow condition of 6 l/s. As can be seen, the plant exhibits a number of antiresonance modes (frequencies at which the magnitude of the frequency response changes abruptly) within the control bandwidth (the main one located around 0.02 rad/s). When exciting the system with a signal with principal frequency components corresponding to those of the antiresonance modes, variations at the system output are very small.

Figure 2.10 shows the theoretical (without taking into account thermal losses) and real frequency response at one of the operating points. As can be seen, the amplitude decreases at certain frequencies but it does not reach zero because the field thermal losses dampen this decrease.

The antiresonance behaviour can be analyzed by considering an approximate simple system which consists of a tube through which fluid is pumped. If this tube is divided into N elements of length Δx metres, considering fluid properties to be constant (in reality they are functions of the temperature) and the thickness of the tube walls equal to zero (or infinite thermal conductivity), by making an energy balance in each of these elements:

$$\frac{\partial(MCT(x,\tau))}{\partial\tau} = I - \rho QC(T(x+\Delta x,\tau) - T(x,\tau))$$

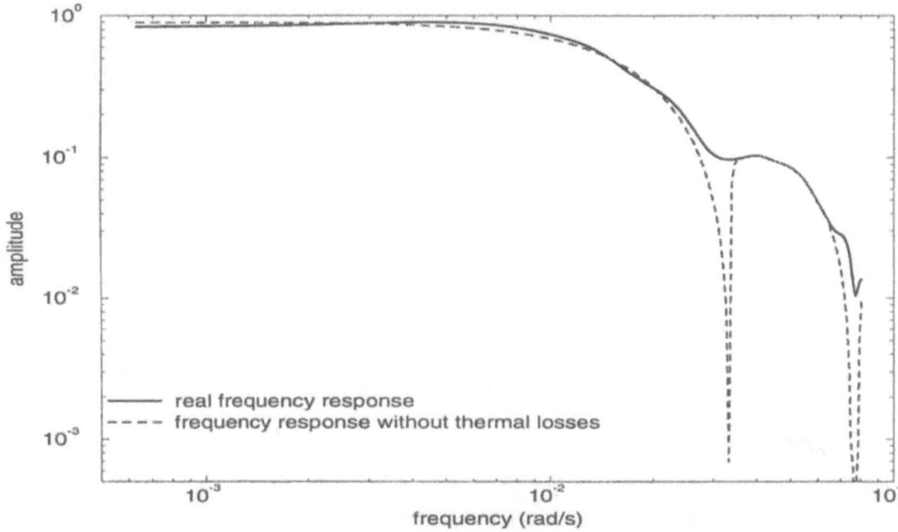

Fig. 2.10. Real and theoretical frequency responses

$$M = \rho A \Delta x$$

where A=area, I=corrected solar radiation, Q=flow, $T(x, \tau)$ is the temperature of oil at a distance x and at time τ, ρ is the density of the oil and C is the thermal capacity.

Considering a small disturbances model around an equilibrium point $(Q_0, T_0(x))$ and neglecting second order terms:

$$T(x, \tau) = T_0(x) + t(x, \tau)$$

$$Q(\tau) = Q_0 + q(\tau)$$

$$\rho AC \frac{\partial t(x, \tau)}{\partial \tau} = -\rho Q_0 C \frac{\partial t(x, \tau)}{\partial x} - \frac{I}{Q_0} q \tag{2.4}$$

If the flow perturbation is made equal to a sine wave:

$$q(\tau) = q' sin(\omega \tau) \tag{2.5}$$

And solving (numerically) equation (2.4) for different values of the excitation frequency, the results shown in Figure 2.11 are obtained. Figure 2.11 shows the amplitude of the sine wave obtained for the temperature of the oil $t(x, \tau)$ along the pipe. As can be seen, the outlet oil temperature has a zero response for a frequency of 0.62 and 1.25 rad/s for a tube of 100 metres. In practice, these frequencies are lower and the drops are attenuated by the field losses and other aspects not considered in equation (2.4).

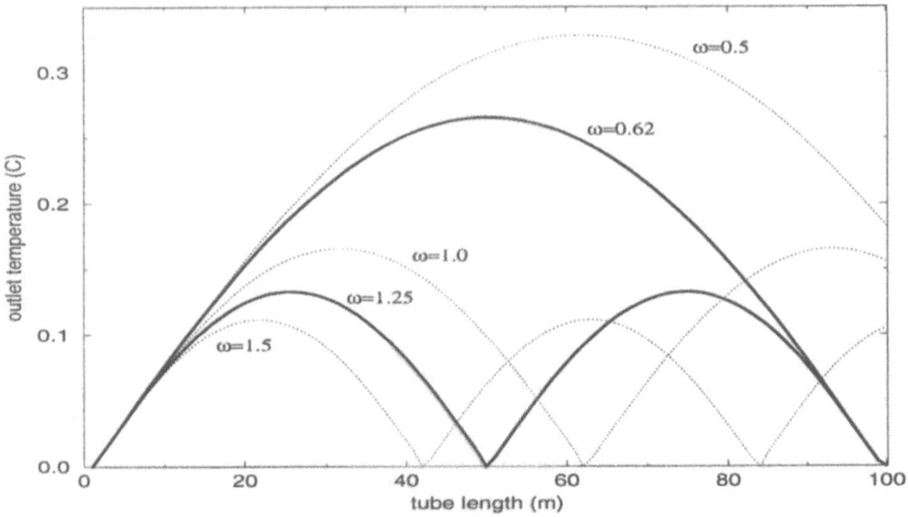

Fig. 2.11. Amplitude of temperature wave in each one of the tube elements

Another way to analyze the antiresonance modes is by considering a sinusoidal solar radiation profile and studying variations produced in the outlet temperature, as done in [81]. In order to summarize the physical mechanism which quantitatively accounts for antiresonance modes, the solar radiation is supposed to be a sinusoid signal to study its effect on the outlet temperature when steady state conditions in the oil flow are considered. The variation in outlet oil temperature Δt_o of one of the loops is a function of the energy accumulated by the fluid when pumped through the tube. If air thermal losses and the thermal capacity of the tube are neglected, the variation in the outlet temperature due to a change in solar radiation can be expressed as the integral of the energy absorbed by the fluid (ΔH) when circulating through the tube:

$$\Delta t_o \approx K \int_0^{\tau_r} \Delta H \cdot d\tau$$

where τ_r is the residence time of the fluid in the field ($\tau_r = L/v$), with L being the tube length and v the fluid velocity. If solar radiation is a sinusoid of frequency ω, when $\omega = \frac{2\pi}{\tau_r}$ (period equal to the residence time), the value of Δt_o is zero and so, no changes are observed in the outlet temperature.

Extending this result, if the period is considered to be 1.5 times the residence time ($\omega = \frac{3\pi}{\tau_r}$), the integral and thus the variations in the output amplitude will be maximum for a determined magnitude of the input signal.

These situations are valid for high frequencies. In general, a minimum will be obtained for $\omega = \frac{2n\pi}{\tau_r}$ and a maximum for $\omega = \frac{2\pi(n+1/2)}{\tau_r}$, with n being a positive integer.

In practice, air thermal losses and the thermal capacity of the tube cannot be neglected and attenuate the variations in the outlet temperature.

From the control viewpoint, the plant can be approximated by simplified linear models when considering the operation around a set point (small signal model). As has been seen in the analysis of the step response, the plant can be approximated by a simplified first order model and a dead time, as in [26], [22]. This is a good approximation if only low frequencies are excited. If this is not the case, as happens when more demands are made on the plant response time, the antiresonance modes (unmodelled dynamics) may give rise to an unacceptable oscillatory behaviour. This is illustrated by Figure 2.12, which shows the response of the plant with a generalized predictive controller designed with a simplified model, with a control weighting factor $\lambda = 5$ and a control horizon of 15. The oscillations show a great amplitude, due to the excitation of the antiresonance modes and because a small value of the weighting factor λ was chosen (see chapter 5).

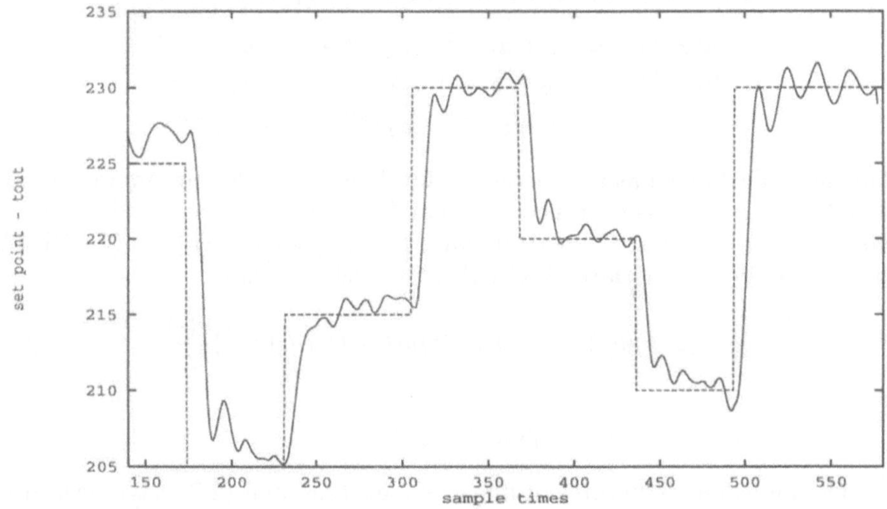

Fig. 2.12. Effects of unmodelled dynamics on the outlet temperature

2.6 Linear plant models

Most single input - single output (SISO) plants, when considering operation around a particular set point and after linearization, can be described in the discrete time domain by linear models based on the concept of transfer function $G(z^{-1}) = B(z^{-1})/A(z^{-1})$ (with z^{-1} being the backward shift operator), relating the system output $y(t)$ with the system input $u(t)$ by:

$$A(z^{-1})y(t) = B(z^{-1})u(t)$$

$$\begin{aligned} A(z^{-1}) &= 1 + a_1 z^{-1} + a_2 z^{-2} + ... + a_{na} z^{-na} \\ B(z^{-1}) &= b_0 + b_1 z^{-1} + b_2 z^{-2} + ... + b_{nb} z^{-nb} \end{aligned}$$

This description is valid both for stable and unstable processes, and has the advantage of needing few parameters to model the system, although it is fundamental to have an *a priori* knowledge of the system, mostly about the order of polynomials $A(z^{-1})$ and $B(z^{-1})$.

If noises and disturbances that can act on the system are taken into account, other types of linear models can be obtained, such as the following:

$$A(z^{-1})y(t) = z^{-d}B(z^{-1})u(t-1) + C(z^{-1})\epsilon(t) \tag{2.6}$$

where $u(t)$ and $y(t)$ are the control and output sequence of the plant, d is the dead time of the system and $\epsilon(t)$ is a zero mean white noise. A, B and C are the following polynomials in the backward shift operator z^{-1}:

$$\begin{aligned} A(z^{-1}) &= 1 + a_1 z^{-1} + a_2 z^{-2} + ... + a_{na} z^{-na} \\ B(z^{-1}) &= b_0 + b_1 z^{-1} + b_2 z^{-2} + ... + b_{nb} z^{-nb} \\ C(z^{-1}) &= 1 + c_1 z^{-1} + a_2 z^{-2} + ... + c_{nc} z^{-nc} \end{aligned}$$

This model is known as a Controller Auto Regressive Moving Average model (CARMA). It has been argued [35] that for many industrial applications in which disturbances are non-stationary, an integrated CARMA (CARIMA) model is more appropriate. A CARIMA model is given by:

$$A(z^{-1})y(t) = z^{-d}B(z^{-1})u(t-1) + C(z^{-1})\frac{\epsilon(t)}{\Delta} \tag{2.7}$$

$$\Delta = 1 - z^{-1}$$

The indicated structures of linear models (2.6) and (2.7) have been used for control design purposes. In the following sections, several linear models obtained from input-output data of the plant are obtained.

2.6.1 Low order linear plant models

Low order models are adequate for many control structures, although their simplicity can produce unacceptable behaviour if fast responses are required from the system, due to the influence of unmodelled dynamics (in this case, the antiresonance modes).

When small changes around a particular set point are considered, most industrial processes can be described by a normally high order linear model. The justification of this affirmation lies in the fact that the majority of the processes are composed of many dynamic elements, usually first order, so that the complete model has an order equal to the number of elements. In fact, each energy or mass storaging element gives rise to a first order element in the model. For instance, a heat exchanger can be modelled by dividing it into segments each one being considered a first order system. The resulting model will have an order equal to the number of segments the tube has been divided into. These high order models are very difficult to cope with in control, but fortunately, the behaviour of such processes can often be modelled by a system with a fundamental time constant and a delay.

Let's consider a process with N first order elements in series, each one having a time constant τ/N. The system transfer function will be given by:

$$G(s) = \frac{1}{(1+\frac{\tau}{N}s)^N}$$

If N is varied from 1 to ∞, the type of response changes from that of a first order system to that of a pure delay (equal to τ). If, as happens in many processes, one of the time constants is greater than the rest, the smaller time constants join to produce a time lag which acts as a pure delay. In this situation, the dynamics are dominated by the major time constant. So, it is possible to approximate the high order model of a complex dynamical system by a first order process with a delay element.

For control design purposes, several linear models which relate changes in outlet oil temperature to changes in the control variable (flow)[2] have been obtained.

These models were obtained using the *reaction curve* method, by producing a step in the oil flow signal (Figure 2.8) at different operating points. The process was approximated by a first order transfer function with a pure delay:

$$G(s) = e^{-s\tau_d}\frac{K}{(1+\tau s)} \tag{2.8}$$

Obviously, this is an approximate model, because the system is nonlinear, but the approximation is made to obtain control schema depending of few parameters to simplify the identification mechanism in the case of adaptive controllers. From the transfer function of equation (2.8) the discrete time transfer function can be obtained taking into account the zero order hold. Two main models have been postulated [103] and are commented below:

Type A model. Since the time delay τ_d is relatively small compared to the fundamental time constant τ, a suitable discrete model can be constructed by choosing the sampling period T to be equal to (or close to) the lowest value of the time delay τ_d (approximately 40 seconds). In this case, the sampling time has to be a multiple of 3 seconds because of the

[2] If the feedforward analyzed in chapter 3 is introduced in series with the plant, the control signal will be the reference temperature to the feedforward controller.

data acquisition system. A sampling time of 39 seconds is therefore used. This sampling time is also adequate considering that the time constant of the process is approximately 6 minutes. This situation corresponds to the operating condition where the flow level is near the highest achievable value. The discrete transfer function model then has the form:

$$G_A(z^{-1}) = z^{-k}\frac{(bz^{-1})}{(1-az^{-1})} \tag{2.9}$$

with $a = e^{-T/\tau}$ and $b = K(1-a)$. At the highest flow level condition $k = 1$.
However, as the flow decreases the system time delay increases and at operating conditions close to the lowest operational flow level to be considered, the delay is approximately doubled, i.e. two sampling periods. For the low flow condition a value of $k = 2$ is appropriate in $G_A(z^{-1})$. The Type A model can represent behaviour at the two extreme conditions of operation when the delay is equivalent to either 1 or 2 sampling periods. It is not designed to cope with intermediate conditions where a non-integer value of k would apply.

Type B model. Another way of coping with the variable time delay condition is to modify the model to the form shown below:

$$G_B(z^{-1}) = z^{-2}\frac{(b_0 + b_1 z^{-1})}{(1-az^{-1})} \tag{2.10}$$

Obviously when $b_1 = 0$ the transfer function is reduced to that of $G_A(z^{-1})$ with $k = 1$, and when $b_0 = 0$ to the case when $k = 2$.
For values of the time delay where τ_d is a non-integer multiple of the sampling period, i.e. $T < \tau_d < 2T$, the factor $(b_0 + b_1 z^{-1})$ acts as a discrete first order Pade approximation to a delay term.

These linear simplified models are based on the step response of the plant. If the frequency response of one of these models is compared to that of the plant or the nonlinear distributed parameter model at an operating point (Figure 2.13), it can be seen that the approximations made in this section are only valid for the low frequency range. The consequences of this fact will be commented in following chapters.

2.6.2 High order linear plant models

In order to account for the antiresonance characteristics of the plant higher order linear models are needed. By using input-output data obtained with PRBS tests, degrees of polynomials $A(z^{-1})$ and $B(z^{-1})$ and that of the delay which best describes the system and minimize the *Akaike's Information Theoretic Criterion* (AIC) [74], were found to be: $n_a = 2$, $n_b = 8$ and $d = 0$.

$$A(z^{-1}) = 1 + a_1 z^{-1} + a_2 z^{-2}$$
$$B(z^{-1}) = b_0 + b_1 z^{-1} + ... + b_8 z^{-8}$$

The value of the coefficients which define the previous polynomials in the backward shift operator z^{-1} were calculated using a least squares algorithm. As an example, using the input-output data with which the frequency response shown in Figure 2.9 was calculated (oil flow around 6 l/s), the values of polynomials $A(z^{-1})$ and $B(z^{-1})$ given by the identification algorithm were:

$$A(z^{-1}) = 1 - 1.5681z^{-1} + 0.5934z^{-2}$$

$$B(z^{-1}) = 0.0612 + 0.0018z^{-1} - 0.0171z^{-2} + 0.0046z^{-3}$$
$$+ 0.0005z^{-4} + 0.0101z^{-5} - 0.0064z^{-6} - 0.015z^{-7} - 0.0156z^{-8}$$

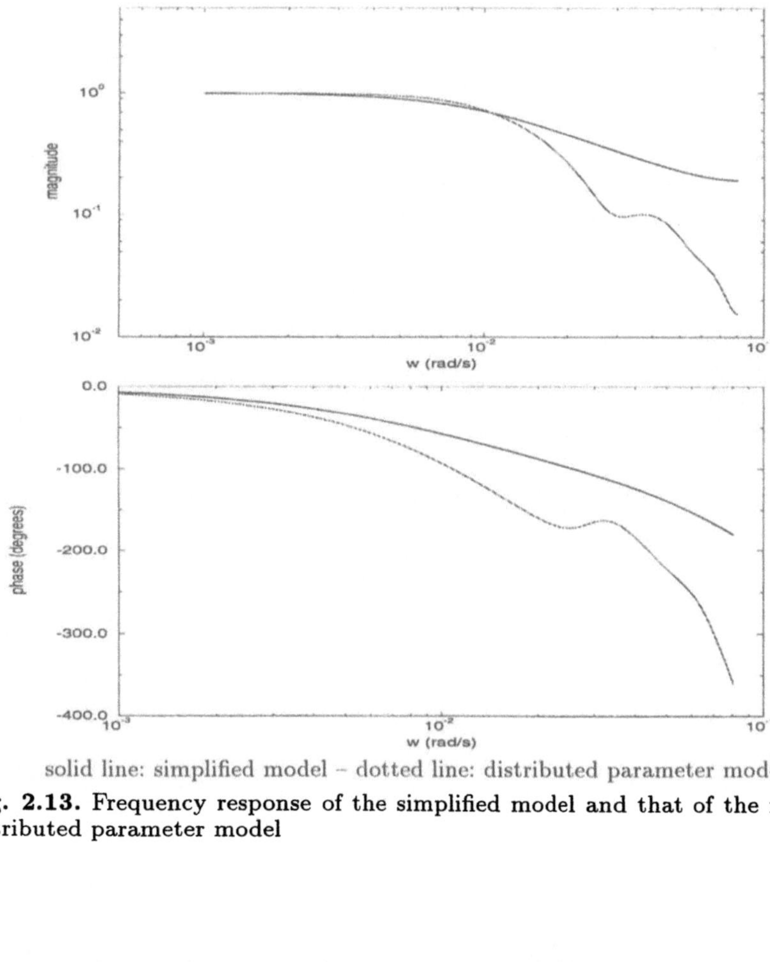

solid line: simplified model – dotted line: distributed parameter model

Fig. 2.13. Frequency response of the simplified model and that of the nonlinear distributed parameter model

CHAPTER 3
BASIC CONTROL SCHEMA

This chapter deals with the development of feedforward controllers, whose introduction within the control loop plays an important role because of their capabilities to reject measurable disturbances which act on the plant, such as solar radiation and inlet oil temperature changes. The chapter also comprises the description of some basic fixed PID control strategies in order to analyze the behaviour obtained with these classical control techniques.

3.1 Feedforward control

Feedforward controllers are extensively used in industry to correct the effect caused by external and measurable disturbances. The disturbances are sensed and used to calculate the value of the manipulated variable required to maintain control at the set point. The first step in designing a feedforward control system is to obtain a mathematical model of how the disturbances affect the process. Using the perturbation (or load) model and the process model, the manipulated variable is computed in order to cancel the effect of the disturbance in the process output.

The offset resulting from modelling errors can be eliminated by adding feedback. This is done by merging a feedforward and a feedback controller. Without feedforward, the control feedback loop is required to change its output to follow all changes of the process load. With feedforward, the feedback controller must only change its output by an amount equal to that which the feedforward system fails to correct.

The distributed solar collector field suffers from changes in the received energy which can be slow, as daily radiation variations, mirror reflectivity changes due to accumulation of dust, etc.; or fast, mainly due to passing clouds and changes in the inlet oil temperature at the starting phase of the power conversion system. These disturbances force the oil flow to change, producing a variable residence time of the fluid within the field.

The fundamental variables which influence the outlet temperature of the field are shown in Figure 3.1. The manipulated variable is the oil flow pumped through the field.

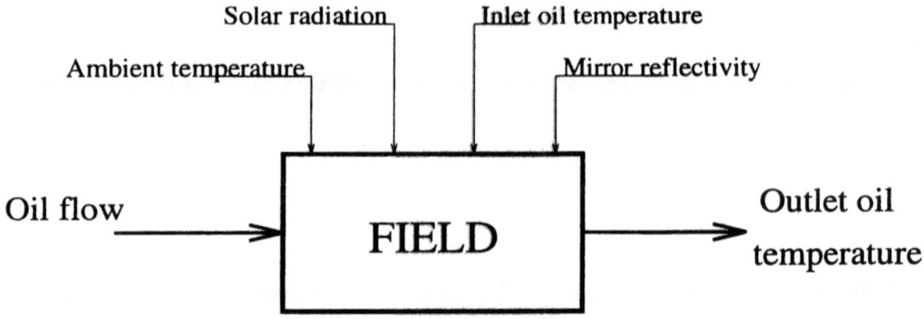

Fig. 3.1. Variables which affect the outlet oil temperature

The outlet temperature of the field depends on other variables such as the solar radiation I and the inlet temperature to the field T_{in}. Mirror reflectivity also has an influence but changes so slowly that it may be considered constant during a test. Hence, dynamically, the outlet temperature T_o can be expressed as a nonlinear function f of oil flow u ($\dot{q}$), solar radiation I and inlet temperature T_{in}. The linearized model is based on partial derivatives (of the change in outlet temperature ΔT_o with respect to changes Δu, ΔI, and ΔT_{in}).

$$T_o = f(u, I, T_{in})$$

$$\Delta T_o = \frac{\partial f}{\partial u}\,\Delta u + \frac{\partial f}{\partial I}\,\Delta I + \frac{\partial f}{\partial T_{in}}\,\Delta T_{in}$$

The partial derivatives can be considered as transfer functions relating the variation in outlet temperature ΔT_o to variations in oil flow Δu, solar radiation ΔI and oil inlet temperature ΔT_{in}, respectively.

The mathematical model which accounts dynamically for these additional influences, even when linearized, is complex. One simple approach which reduces the complexity of the model is to make use of the known operating characteristics of the field. Studies have been made which incorporated approximate transfer functions in the feedforward term but the benefits provided were not considered to justify the added complexity or uncertainty involved. A feedforward term has been developed [26] based directly on the steady state energy balance relationship given below.

By using the equation (2.1), which describes the concentrated model of the field analyzed in chapter 2, by considering steady state conditions and by using experimental data, it is possible to derive a correlation for the oil flow as function of the inlet and outlet oil temperatures, solar radiation, mirror reflectivity and ambient temperature. Once this equation is obtained, it can be used as a feedforward action in the control system.

From this basic structure of feedforward control, two simple approaches are presented in order to compensate for effects of the inlet oil temperature and solar radiation on the outlet temperature. The first one employs a parallel feedforward term, and the second introduces a series compensation element into the loop. In both cases, the radiation and inlet oil temperature serve to directly adjust the oil flow to the value calculated to maintain the outlet temperature at the desired level. This restricts the outlet temperature excursions, which is desirable from the control viewpoint, and ensures that the outlet temperature is predominantly a function of the oil flow, which is the manipulated variable.

Taking into account the previous basic considerations, one has:

$$(T_o - T_{in})u = \frac{1}{Pc_p}(n_o SI - H_l(T_m - T_a))$$

with $u = \dot{q}$ being the flow of the pump . This last equation can be approximated by [29]:

$$(T_o - T_{in})u = 0.7869I - 0.485(T_o - 151.5) - 80.7$$

where the constants that appear in the equation have been determined experimentally from the basic formulation [99].

The feedforward controllers developed from this equation have proved to be effective in many of the tests performed at the plant. As indicated in [53], in most cases it is sufficient to use a feedforward constituted by a simple proportional term to compensate for measurable disturbances.

3.1.1 Parallel feedforward compensation

The basic structure of a parallel feedforward controller is shown in Figure 3.2, where variable u_{ff} is the flow calculated to provide the desired outlet temperature (T_r) for the prevailing values of I and T_{in}. The calculation employed is:

$$u_{ff} = \frac{0.7869I - 0.485(T_r - 151.5) - 80.7}{T_r - T_{in}}$$

When variations occur in I and T_{in}, the signal u_{ff} is changed to a value which, in steady state, would maintain the desired outlet temperature at the desired value. This feedforward element serves to significantly reduce the dynamic variations in T_o due to changes in I and T_{in}, and provides quick response to reference temperature changes. In terms of estimation without the added signal u_{ff}, the outlet temperature variations lead to significant and unnecessary adjustments being made to the estimated parameters due to changes in inlet temperature and solar radiation (unaccounted for in the SISO linear models described in chapter 2 and used by many control algorithms). This tendency, although not completely eliminated, is significantly reduced by including u_{ff}. The added signal can be considered as a feedforward term which helps to maintain the desired output temperature when measurable disturbances occur.

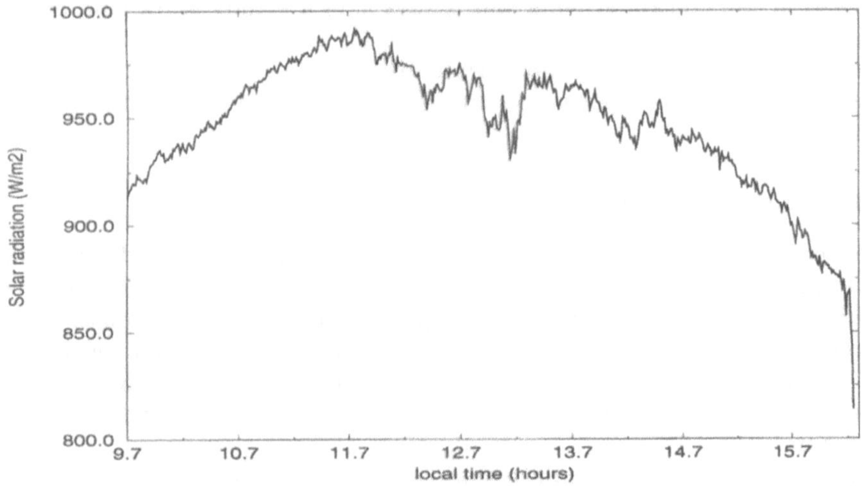

Fig. 3.4. Solar radiation during a test with basic control schema

As can be seen good results have been obtained in both cases, in spite of disturbances acting during the operation, the main design objective being achieved, that is to obtain a response with small overshoot.

In principle it is possible to use either of the two feedforward structures, but with the parallel way of compensation, injected disturbances for estimation purposes cannot be applied via reference temperature variations, as they have a direct effect on u_{ff}, and therefore on the input flow, which is not accounted for by the flow signal used in the estimation algorithm.

The use of series feedforward compensation has been considered to be more advantageous since the whole plant and the controller become a system that approaches a linear one (at least in terms of small variations around an operating point), which provides large benefits when using an identification mechanism. When variations occur in I or T_{in}, the series feedforward calculates the value of flow needed to maintain the desired outlet temperature. By reducing the effect on the outlet temperature of disturbances in variables unaccounted for by the model, its validity is preserved and unwanted and unnecessary adjustments in the estimation parameters are reduced, when a self-tuning controller is incorporated.

The series feedforward can be considered as part of the plant and the input signal will then be the reference temperature for the feedforward controller. By doing this, the resulting model will have a gain of approximately unity, and the static nonlinearities will be cancelled. The output of the feedforward controller is the reference signal for the oil flow fixed PID controller.

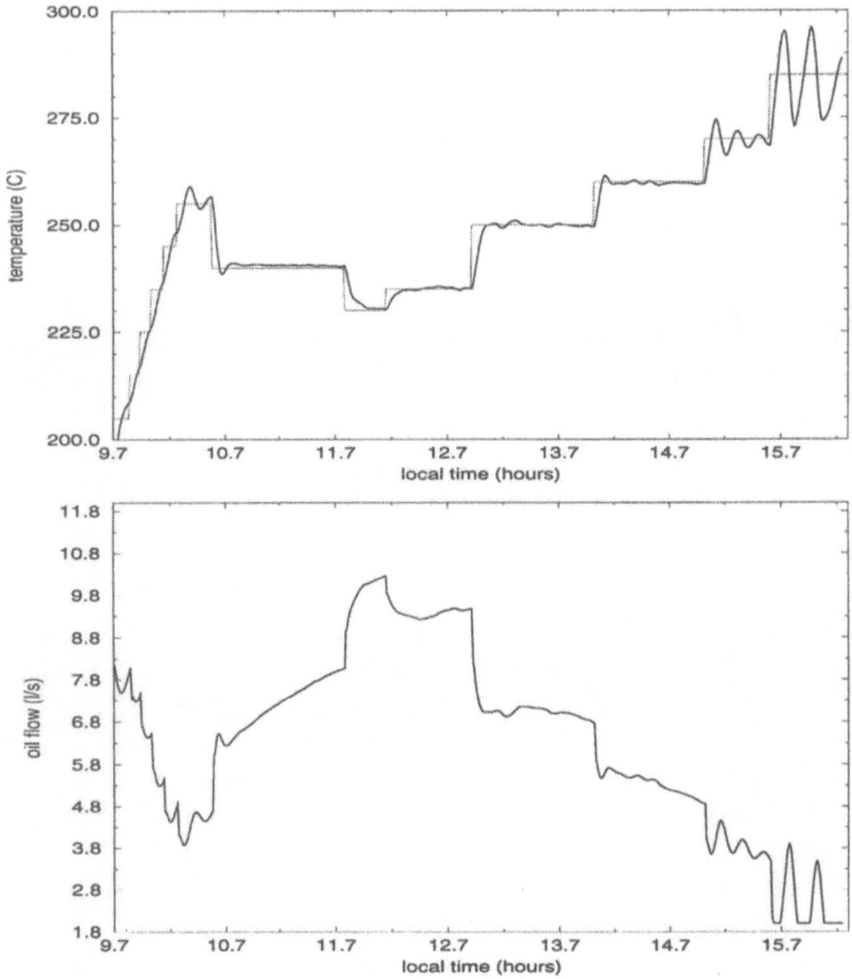

Fig. 3.5. Response without feedforward compensation

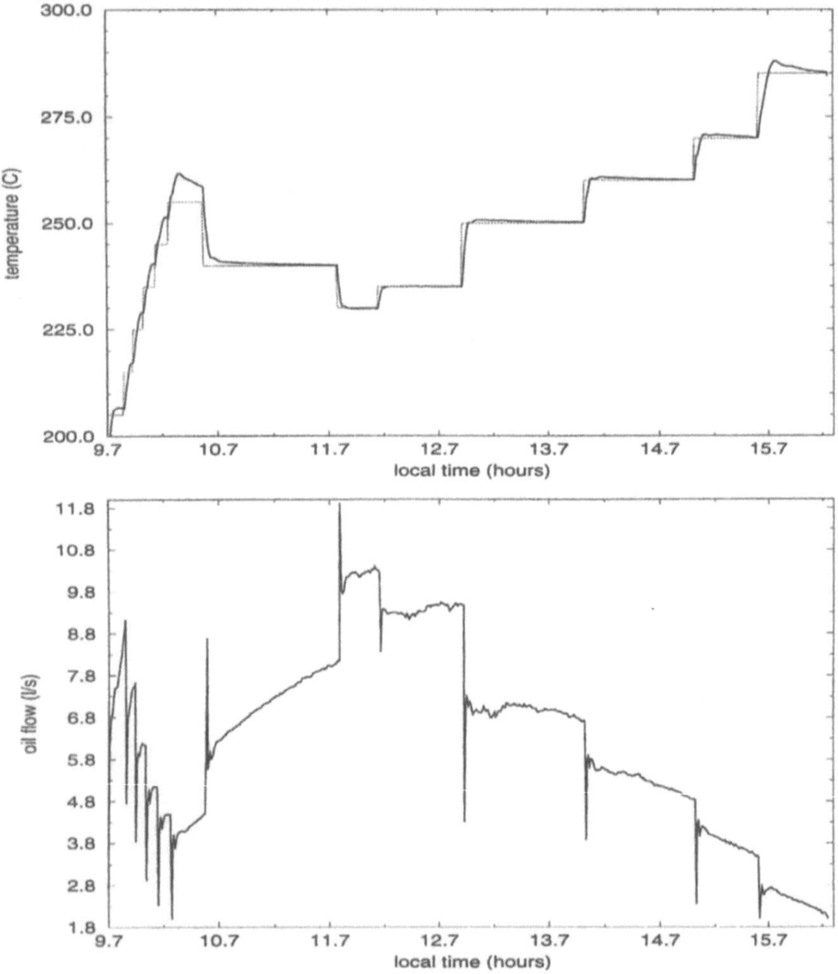

Fig. 3.6. Response with parallel feedforward compensation

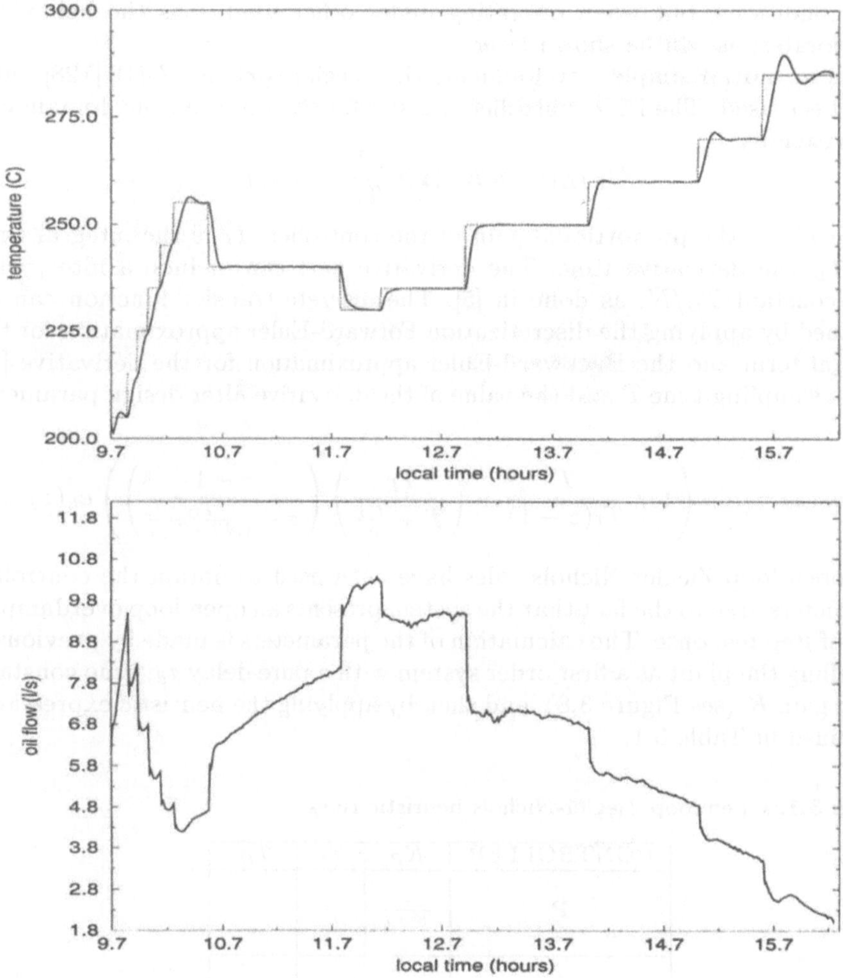

Fig. 3.7. Response with series feedforward compensation

3.2 Fixed Ziegler-Nichols rule based PID controllers

As a prior step to the development of advanced control strategies, different classical PID controllers were designed and tested at the plant (simulations have been shown in the previous section). With this kind of fixed controllers, good behaviour can be achieved when operating near the design or nominal flow conditions, but when operating under other conditions the behaviour deteriorates, as will be shown later.

As an initial simple development, the Ziegler-Nichols (Z-N) [128] rules have been used. The PID controller transfer function in the time domain can be written as:

$$G_{PID}(s) = K_P(1 + \frac{1}{T_I s} + T_D s)$$

where K_P is the proportional gain of the controller, T_I is the integral time and T_D the derivative time. The derivative part can include a filter, with time constant T_D/N, as done in [5]. The discrete transfer function can be obtained by applying the discretization Forward-Euler approximation for the integral term and the Backward-Euler approximation for the derivative [5], with a sampling time T and the value of the derivative filter design parameter $N = 5$.

$$u_k(z) = K_P\left(1 + \frac{T}{T_I(z-1)} + \left(\frac{T_D}{T + \frac{T_D}{N}}\right)\left(\frac{z-1}{z - \frac{T_D}{(NT+T_D)}}\right)\right) e_k(z)$$

The open loop Ziegler-Nichols rules have been used to obtain the controller parameters, due to the fact that the system presents an open loop overdamped kind of step response. The calculation of the parameters is made by previously modelling the plant as a first order system with a pure delay τ_d, time constant τ and gain K (see Figure 3.8), and then by applying the heuristic expressions contained in Table 3.1.

Table 3.1. Open loop Ziegler-Nichols heuristic rules

CONTROLLER	K_P	T_I	T_D
P	$\frac{\tau}{K \cdot \tau_d}$	–	–
PI	$\frac{0.9\tau}{K \cdot \tau_d}$	$3\tau_d$	–
PID	$\frac{1.2\tau}{K \cdot \tau_d}$	$2\tau_d$	$0.5\tau_d$

The controllers designed by this method usually provide fast responses with overshoots of about 25%. Thus, it is usual to reduce the value of the proportional gain in order to avoid highly underdamped responses.

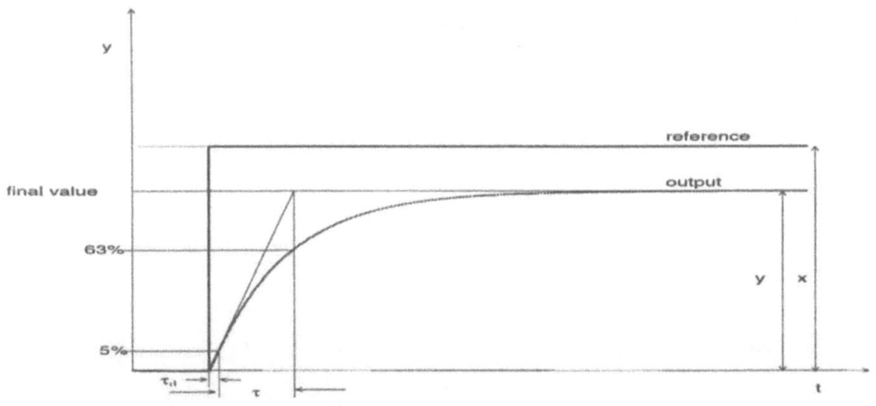

Fig. 3.8. First order system response characterization

3.3 Backup controller

Operation with a fixed PI backup controller ($K_P = 1.56$, $T_I = 675$ s.) including series feedforward compensation is shown in Figure 3.9. Such a controller was used as an auxiliary controller in starting or emergence procedures. The behaviour of the controller is acceptable at the operating point for which it was designed (medium oil flow), but it deteriorates under other operating conditions.

Due to the fact that the environmental conditions of one specific test cannot be reproduced at the actual plant, in order to analyze and compare the behaviour of the proposed control schema, data obtained when operating with this fixed PI backup controller (solar radiation, inlet oil temperature, ambient temperature, mirror reflectivity, etc.) have also been used as a reference test throughout the text. The evolution of the first two mentioned variables is shown in Figure 3.10

3.4 Fine-tuned PID controller

The PID controller designed by the Z-N rules can be improved by modifying the controller parameters after the first tests carried out at the plant. Several years' experience and the availability of the nonlinear distributed parameter model allowed us to tune a fixed PID controller to operate at medium flows (between 6 l/s and 7 l/s). As will be seen, this controller has been tuned so that the responses obtained were as fast as many advanced control strategies based on low order models around the nominal design point. At operating points other than the nominal one, the behaviour deteriorates, producing high oscillations when operating at low flows.

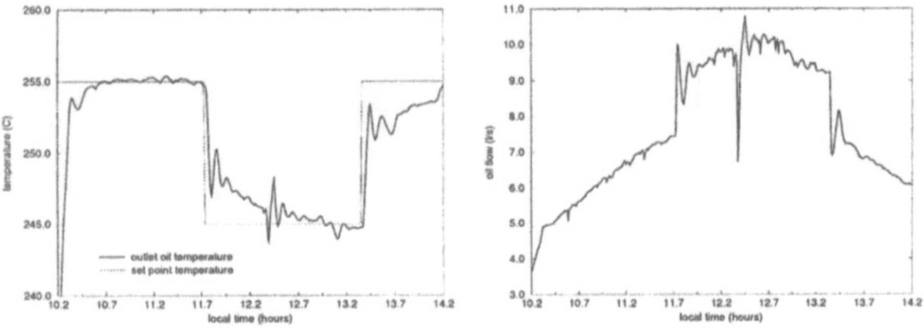

set point and outlet oil temperature (°C) oil flow (l/s)

Fig. 3.9. Test with the backup PI controller (26/11/90)

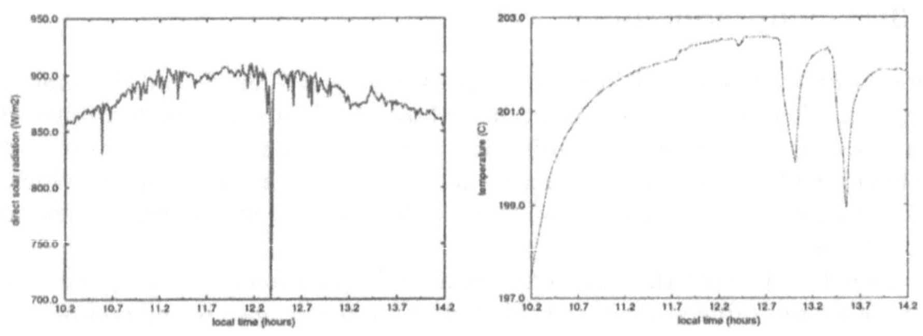

solar radiation in the reference test (W/m²) inlet oil temperature in the reference test (°C)

Fig. 3.10. Solar radiation and inlet oil temperature conditions in the reference tests

Figure 3.11 shows the result obtained with the initial PID controller design based on Ziegler-Nichols rules ($K_P = 3$, $T_I = 120$ s., $T_D = T_I/4$ s.) and that ($K_P = 0.5$, $T_I = 75$ s., $T_D = T_I/4$ s.) obtained after modifications when operating around the nominal set point taking into account various tests performed under those oil flow conditions, and using the nonlinear distributed parameter model of the field.

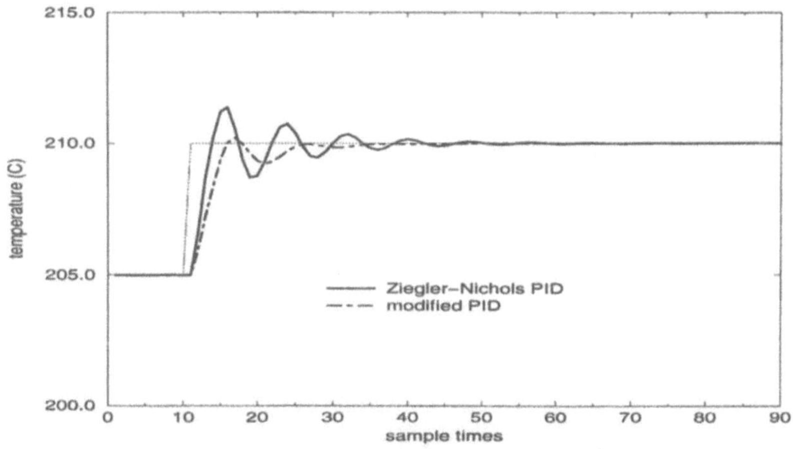

Fig. 3.11. Design of the fixed PID controller

As an illustrative example, Figure 3.12 shows the results obtained when controlling the plant with the described PID controller during an operation covering different working points. Figure 3.13 shows the direct solar radiation conditions and the oil flow pumped through the field in this test.

As can be seen, the results obtained when operating around medium flows (6-7 l/s) are excellent, but the behaviour of the system controlled deteriorates under other operating conditions (high or low flow). Notice that in conditions of low oil flow (at the end of operation), a large overshoot and settling time appear in the response.

By looking at the responses obtained from the system under different operating conditions the following comments can be made:

- Under low flow conditions, the controlled systems tend to have an oscillatory response, mainly due to the following reasons:
  - At low flow, the residence time of the fluid through the field is greater than with higher flow levels, the antiresonance modes lying at lower frequencies. This fact can be observed in the plant Bode plots in chapter 5, in which, the first antiresonance mode lies at lower frequencies as the flow decreases.
  - As flow decreases, the delay of the system increases and the relative robustness margins decrease.
  - At low flow conditions, the system is more sensitive to pumped flow variations (the system gain increases as the flow decreases). The curves showing the response of different controllers show how, at low flow conditions, a variation of about 0.15 l/s produces a variation in the outlet

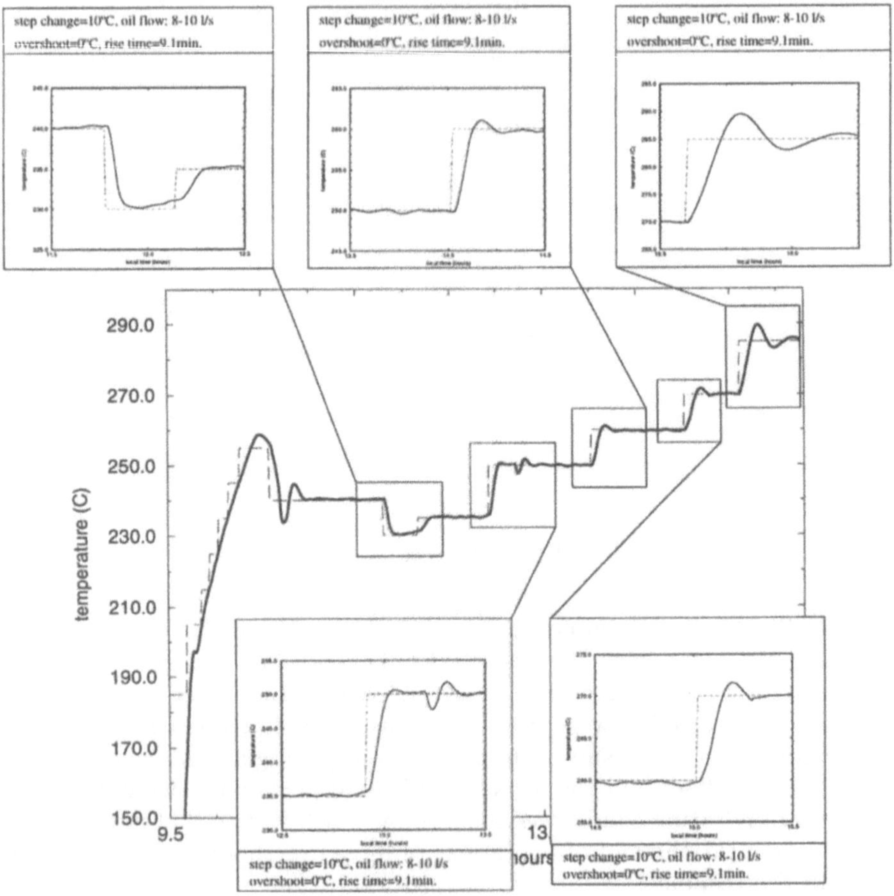

Fig. 3.12. Test with the fine-tuned fixed PID controller (27/03/96)

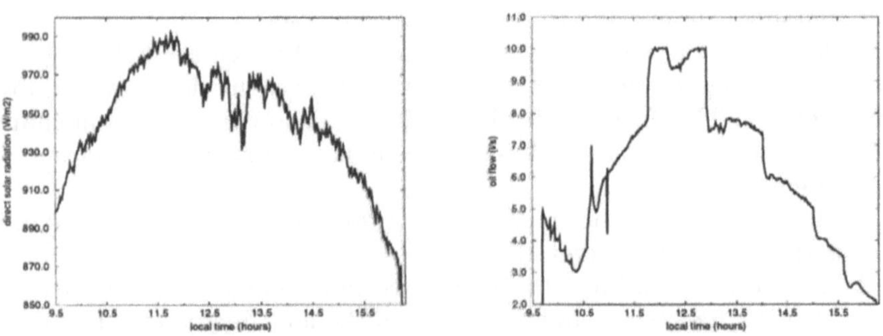

direct solar radiation (W/m^2) oil flow (l/s)

Fig. 3.13. Test with the fine-tuned fixed PID controller (27/03/96) oil temperature of 5°C, while at medium and high flow conditions, variations of 0.6 and 1.6 l/s are required to obtain the same variation (5°C) in the outlet temperature. It is important to point out that this comment does not contradict the supposition made throughout the book that the system with series feedforward has an almost unitary steady state gain and mainly dependent on steady state conditions of direct solar radiation and inlet oil temperature. This is due to the formulation relating the flow to the reference temperature given to the feedforward controller. Under the conditions of the proposed simulation, Figure 3.14 shows the mentioned relationship.

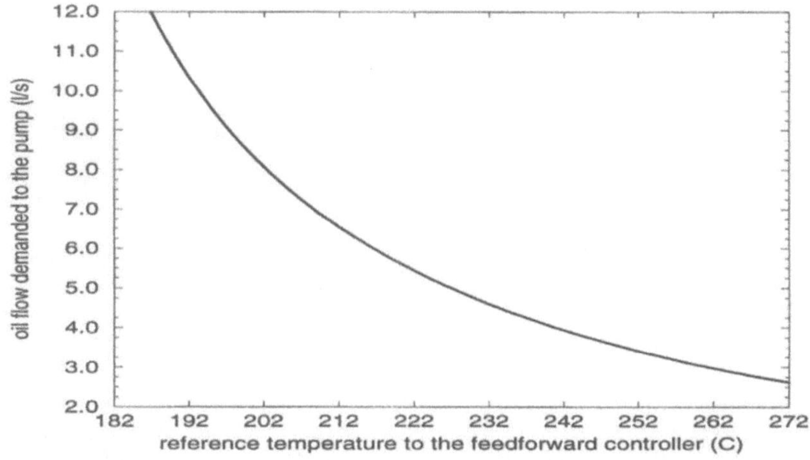

Fig. 3.14. Demanded oil flow *versus* feedforward reference temperature

As can be seen, the same increment in the oil flow signal produces greater feedforward reference temperature changes (and thus greater changes in the outlet temperature as the plant is operated in closed loop) at low flow conditions than at high flow levels. This nonlinear relationship is responsible for obtaining an almost unitary steady state gain, but makes this gain dependent on the operating conditions.

- At high flow level conditions the delay decreases. The antiresonance modes are not significant in this case as these modes lie at high frequencies.
- From the viewpoint of disturbance rejection capabilities, it can be seen that a great amount of the contribution needed to compensate for disturbances is provided by the series feedforward controller. The rest of the contribution is provided by the feedback controllers (especially when coping with large low-frequency-gain controllers).

The conclusion that can be drawn from this study is that a fixed PID controller does not work appropriately at all operating points and more advanced controllers should be designed in order to take into account the variations in process dynamics and the antiresonance modes.

CHAPTER 4

BASIC STRUCTURES OF ADAPTIVE CONTROL

This chapter deals with the description of the parameter estimation algorithm used in all the developed adaptive control schema, including the supervisory mechanisms needed for assuring some degree of robustness and performance. The chapter also includes the description of some adaptive PID control strategies.

The main objective of the control system is to maintain the outlet oil temperature as close as possible to the reference temperature. Since the solar radiation cannot be adjusted, this can only be achieved by the adjustment of the flow of oil, and the daily solar power cycle characteristic is such that the oil flow has to change substantially during operation. This leads to significant variations in the dynamic characteristics of the field such as response rate and time delay which cause difficulties in obtaining adequate performance over the operating range with a fixed parameter controller.

Self-tuning control offers an approach by which controller parameters are adjusted during operation to compensate for changes in the dynamic characteristics of the field and thereby to maintain the desired control performance.

The adaptive control scheme used in several of the control strategies presented in this book corresponds to the *self-tuning regulator*, which mainly consists of calculating controller parameters supposing that the system model parameters are those obtained by an identification algorithm. The adaptive control scheme can be seen in Figure 4.1. At each sampling time, the adaptive control strategy consists of:

1. Estimate the linear model parameters using input-output data from the process.
2. Adjust controller parameters (this does not necessarily have to be done each sampling time).
3. Calculate the control signal.
4. Supervise the correct behaviour of the controlled system.

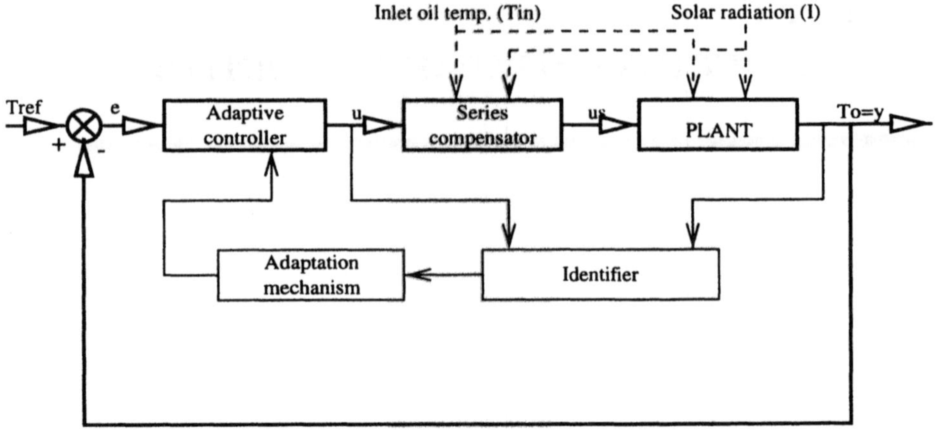

Fig. 4.1. General adaptive control scheme

4.1 Parameter estimation algorithm

This section describes the identification algorithm used for all adaptive control schemes described in the book

4.1.1 Parametric identification

System identification can be defined as the process of obtaining a model for the behaviour of a plant based on plant input and output data. If a particular model structure is assumed, the identification problem is reduced to obtaining the parameters of the model. The usual way of doing this is by optimizing a function that measures how well the model, with a particular set of parameters, fits the existing input-output data. When process variables are perturbed by noise of a stochastic nature, the identification problem is usually interpreted as a parameter estimation problem. This problem has been extensively studied in literature for the case of processes which are linear on the parameters to be estimated, and perturbed with a white noise.

If we suppose that the system can be modelled as a stable, linear, time invariant process (LTI), single input - single output (SISO), it can be described by a difference equation of the form:

$$y(k) + a_1 y(k-1) + \ldots + a_n y(k-n) = b_1 u(k-d-1) + b_2 u(k-d-2) + \ldots$$
$$\ldots + b_n u(k-d-n) + v(k) + c_1 v(k-1) + c_2 v(k-2) + \ldots + c_n v(k-n) \quad (4.1)$$

which expressed in vectorial form gives: $y(k) = \boldsymbol{\varphi}^T(k)\boldsymbol{\theta} + v(k)$, where:

$$\begin{aligned} \varphi^T(k) &= [-y(k-1), -y(k-2), \ldots, -y(k-n), u(k-d-1), u(k-d-2), \\ &\quad \ldots, u(k-d-n), v(k-1), v(k-2), \ldots, v(k-n)] \end{aligned}$$

with d being the pure input-output delay.

$$\boldsymbol{\theta}^T = [a_1, a_2, ..., a_n, b_1, b_2, ..., b_n, c_1, c_2, ..., c_n]$$

$$\begin{aligned} u(k) &= U(k) - U_\infty \\ y(k) &= Y(k) - Y_\infty \end{aligned}$$

$U(k)$ and $Y(k)$ being the system input and output signals in the k_{th} instant, U_∞ the mean value of the input signal, Y_∞ is the mean value of the output signal and $v(k)$ is a noise signal statistically independent and stationary with normal distribution and zero mean. If we assume that disturbances can be modelled by an Auto Regressive Moving Average (ARMA) process with the same characteristic polynomial as that of the system, the discrete transfer function of the system is given by:

$$y(z^{-1}) = \frac{B(z^{-1})}{A(z^{-1})} z^{-d} u(z^{-1}) + \frac{C(z^{-1})}{A(z^{-1})} v(z^{-1}) \tag{4.2}$$

where:

$$\begin{aligned} A(z^{-1}) &= 1 + a_1 z^{-1} + a_2 z^{-2} + ... + a_n z^{-n} \\ B(z^{-1}) &= b_1 z^{-1} + b_2 z^{-2} + ... + b_n z^{-n} \\ C(z^{-1}) &= 1 + c_1 z^{-1} + c_2 z^{-2} + ... + c_n z^{-n} \end{aligned}$$

the first quotient $B(z^{-1})/A(z^{-1})$ representing the plant model and the second $C(z^{-1})/A(z^{-1})$ the disturbances model.

If we suppose $C(z^{-1}) = 1$, $v(k) = e(k)$, and that this residual error is uncorrelated to the elements of $\varphi(k)$, zero mean with normal distribution, the system model results as:

$$y(k) = \varphi^T(k)\boldsymbol{\theta}(k) + e(k) \tag{4.3}$$

where:

$$\begin{aligned} \varphi^T(k) = & [-y(k-1), -y(k-2), ..., -y(k-n), u(k-d-1), u(k-d-2), \\ & ..., u(k-d-n)] \end{aligned}$$

with d being the pure input-output delay.

$$\boldsymbol{\theta}^T = [a_1, a_2, ..., a_n, b_1, b_2, ..., b_n]$$

The first term of the second member of equation (4.3) can be interpreted as the one-step prediction $\hat{y}(k/k-1)$ of the output $y(k)$ with data available at time $k-1$, and so, the error is the difference between the real output and its prediction: $e(k) = y(k) - \hat{y}(k/k-1)$.

Suppose that we have a set of measurements $y(k), y(k+1), ..., y(k+N)$. The problem is solved by finding the set of parameters $\boldsymbol{\theta}$ which minimizes the sum of the weighted square errors:

$$J(\boldsymbol{\theta}) = \sum_{j=0}^{N} \lambda_j e(k+j)^2$$

which can be written as

$$J(\boldsymbol{\theta}) = (\mathbf{Y} - \boldsymbol{\Phi}\boldsymbol{\theta})^T \boldsymbol{\Lambda} (\mathbf{Y} - \boldsymbol{\Phi}\boldsymbol{\theta})$$

where

$$\boldsymbol{\Lambda} = \begin{bmatrix} \lambda_0 & 0 & \dots & 0 \\ 0 & \lambda_1 & \dots & \vdots \\ \vdots & \dots & \ddots & 0 \\ 0 & \dots & 0 & \lambda_N \end{bmatrix} ; \mathbf{Y} = \begin{bmatrix} y(k) \\ y(k+1) \\ \dots \\ y(k+N) \end{bmatrix} ; \boldsymbol{\Phi} = \begin{bmatrix} \varphi^T(k) \\ \varphi^T(k+1) \\ \dots \\ \varphi^T(k+N) \end{bmatrix}$$

The weight factors λ_j are used to penalize the errors of the different measurements taken. If the process is time variant, these factors can be chosen in order to give more weight to the latest measurements. Usually λ_i is chosen to be $\lambda_i = c^{N-k}$ where c is a constant smaller than one known as the forgetting factor.

The minimum can be found by solving

$$\frac{dJ(\boldsymbol{\theta})}{d\boldsymbol{\theta}} = 0,$$

which turns out to be:

$$\boldsymbol{\theta} = (\boldsymbol{\Phi}^T \boldsymbol{\Lambda} \boldsymbol{\Phi})^{-1} \boldsymbol{\Phi}^T \boldsymbol{\Lambda} \mathbf{Y} \tag{4.4}$$

4.1.2 Recursive least squares identification algorithm (RLS)

As the plant is a nonlinear system which is going to be identified as a linear one, parameters of the linear model will change with the plant operating point and the linear model will be time variant. Equation (4.4) can be solved recursively by the well known recursive least squares identification algorithm [46], [6]. In the RLS algorithm, the estimation of parameters ($\boldsymbol{\theta}$) is given by the values of the previous estimation corrected by a linear term in the error between the output and the prediction, being $\mathbf{L}$ the gain of the correction.

The RLS method has been chosen with a variable forgetting factor in order to reduce the identifier memory and to avoid the identifier gain reaching zero. This factor is made equal to 1 if the trace of the covariance matrix exceeds a certain value. Also, the trace must not be less than other *a priori* selected value. The algorithm is well known and consists of the following steps in the recursive version [104]:

1. Select initial values of $\mathbf{P}(k)$, $\hat{\boldsymbol{\theta}}(k)$ and $c(k)$, (covariance matrix, estimated parameters vector and forgetting factor respectively).

2. Read the new values of $y(k+1)$ and $u(k+1)$, (identifier inputs).

3. Calculate the *a priori* residual error:

$$e(k+1) = y(k+1) - \varphi^T(k+1)\hat{\theta}(k)$$

4. Calculate $\mathbf{L}(k+1)$ given by expression:

$$\mathbf{L}(k+1) = \frac{\mathbf{P}(k)\varphi(k+1)}{c(k) + \varphi^T(k+1)\mathbf{P}(k)\varphi(k+1)}$$

5. Calculate the new estimated parameters given by:

$$\hat{\theta}(k+1) = \hat{\theta}(k) + \mathbf{L}(k+1)e(k+1)$$

6. Calculate the new forgetting factor $c(k+1)$.

$$c(k+1) = 1 - (1 - \varphi^T(k+1)\mathbf{L}(k+1))\frac{e(k+1)^2}{S_o}$$

If $c(k+1) < c_{min}$ then $c(k+1) = c_{min}$
If $c(k+1) > 1$ then $c(k+1) = 1$
Parameter S_o must be *a priori* known [45] (S_o is related to the sum of the square errors).

7. Actualize the covariance matrix.

$$\mathbf{P}'(k+1) = (\mathbf{I} - \mathbf{L}(k+1)\varphi^T(k+1))\frac{\mathbf{P}(k)}{c(k)}$$

$$\text{If} \quad tr(\frac{\mathbf{P}'(k+1)}{c(k+1)}) > tr_{max} \quad \text{then} \quad c(k+1) = 1$$

$$\mathbf{P}(k+1) = \frac{\mathbf{P}'(k+1)}{c(k+1)}$$

8. Make $k = k+1$ and return to step 2.

Values of tr_{max} and the initial forgetting factor will depend on the system and the desired convergence speed and stability. Notice that for a value of the forgetting factor equal to unity, the covariance matrix decreases monotonously, allowing the identifier gain to be zero. Moreover, if the operating point is fixed, the product $\mathbf{P}(k)\varphi(k)$ can be zero, and therefore $\mathbf{P}(k+1) = \mathbf{P}(k)/c(k)$, which for $c(k) < 1$, will make $\mathbf{P}(k)$ to increase excessively, making the identifier very sensitive to any change. These are the reasons which justify the selection of a variable forgetting factor in the identification algorithm, so that, if the trace of the covariance matrix exceeds a determined value, the forgetting factor is fixed $c(k) = 1$. Similarly, the trace of the covariance matrix is not allowed to decrease below a certain prefixed value. This is done by a supervisory level of the identification process and by adding some heuristic considerations.

Another problem which can appear when applying the algorithm is that expression:

$$\mathbf{P}(k+1) = (\mathbf{I} - \mathbf{L}(k+1)\varphi^T(k+1))\frac{\mathbf{P}(k)}{c(k)} \tag{4.5}$$

is ill conditioned [32], and the implementation using these equations often leads the matrix $\mathbf{P}$ to be negative definite, obtaining results far from reality. As $\mathbf{P}$ must be positive definite, a solution is to factorize $\mathbf{P}$ as $\mathbf{P}(k+1) = \mathbf{U}(k+1)\mathbf{D}(k+1)\mathbf{U}^T(k+1)$ (UDU factorization [16]), where $\mathbf{D}$ is a diagonal matrix and $\mathbf{U}$ is an upper triangular matrix with its diagonal elements equal to unity. If we define $\bar{\mathbf{U}}, \bar{\mathbf{D}}$ as $\mathbf{U}(k), \mathbf{D}(k)$ respectively, and they are introduced in expression (4.5):

$$\begin{aligned}\mathbf{UDU}^T(k+1) &= \frac{\bar{\mathbf{U}}\bar{\mathbf{D}}\bar{\mathbf{U}}^T}{c(k)} - \frac{\bar{\mathbf{U}}\bar{\mathbf{D}}\bar{\mathbf{U}}^T\varphi(k+1)\varphi^T(k+1)}{c(k)+\varphi^T(k+1)\bar{\mathbf{U}}\bar{\mathbf{D}}\bar{\mathbf{U}}^T\varphi(k+1)}\frac{\bar{\mathbf{U}}\bar{\mathbf{D}}\bar{\mathbf{U}}^T}{c(k)} \\ &= \bar{\mathbf{U}}\left[\frac{\bar{\mathbf{D}}}{c(k)} - \frac{\bar{\mathbf{D}}\bar{\mathbf{U}}^T\varphi(k+1)\varphi^T(k+1)}{c(k)+\varphi^T(k+1)\bar{\mathbf{U}}\bar{\mathbf{D}}\bar{\mathbf{U}}^T\varphi(k+1)}\frac{\bar{\mathbf{U}}\bar{\mathbf{D}}}{c(k)}\right]\bar{\mathbf{U}}^T\end{aligned}$$

The following vectors can be defined: $\mathbf{f} = \bar{\mathbf{U}}^T\varphi, \mathbf{g} = \bar{\mathbf{D}}\mathbf{f}$, leading to:

$$\mathbf{UDU}^T(k+1) = \bar{\mathbf{U}}\left[\frac{1}{c(k)}\left(\bar{\mathbf{D}} - \frac{\mathbf{gg}^T}{c(k)+\mathbf{f}^T\mathbf{g}}\right)\right]\bar{\mathbf{U}}^T$$

If the term between brackets is broken into $\mathbf{U}_n\mathbf{D}_n\mathbf{U}_n^T$, then the new value of the terms is given by:

$$\mathbf{UDU}^T(k+1) = \bar{\mathbf{U}}\mathbf{U}_n\mathbf{D}_n\mathbf{U}_n^T\bar{\mathbf{U}}^T \rightarrow \mathbf{D}(k+1) = \mathbf{D}_n, \quad \mathbf{U}(k+1) = \bar{\mathbf{U}}\mathbf{U}_n$$

4.1.3 Stability and robustness in the identification

Stability and robustness studies of the identification algorithms try to analyze the causes which lead these algorithms to converge onto wrong values and to study their sensitivity in spite of disturbances in the dynamic information entering the identifier. In order to obtain stability increases, several modifications have been carried out in the algorithm originally designed for the ideal case. An increase in the robustness of the adaptive control schema can be achieved by using data prefiltering for identification purposes to obtain better behaviour of the adaptive controller and to assure the validity of the assumed low order model. An excellent development and overview of all these techniques can be found in [87] from which we here summarize simple modifications that can be performed to increase the robustness of the RLS identification mechanism.

Let's consider a SISO process of the form:

$$A^0(z)y(t) = B^0(z)u(t) + A^0(z)[\chi(t) + \psi(t)] \tag{4.6}$$

where $\chi(t)$ and $\psi(t)$ indicate predictable and stochastic disturbances respectively. In particularly, it is supposed that $\Pi(z)\chi(t) = 0$ for a monic polynomial $\Pi(z)$ with all its roots of multiplicity one on the unitary circle. To account for constant disturbances, it is also supposed that $\Pi(1) = 0$. Equation (4.6) can be written as

$$A_0(z)y(t) = B_0(z)u(t) + A_0(z)\psi(t) \tag{4.7}$$

where $A_0(z) = A^0(z)\Pi(z)$ y $B_0(z) = B^0(z)\Pi(z)$. Suppose that the plant is represented by a low order model as

$$A(z)y(t) = B(z)u(t) + n(t) \tag{4.8}$$

where

$$A_0(z) = A(z)A^u(z) \qquad B_0(z) = B(z)B^u(z) \tag{4.9}$$

and

$$n(t) = \frac{B(z)[B^u(z) - A^u(z)]}{A^u(z)}u(t) + A(z)\psi(t)$$

$A^u(z)$ and $B^u(z)$ being monic polynomials and superindex u indicating *unmodelled*. Notice that, as $B^u(z)$ is monic, the modelled part retains the system delay. The idea here is to identify the parameters of the reduced model (4.8), that is, polynomials $A(z)$ and $B(z)$, by a standard identification mechanism and for controller design purposes to utilize the estimated values of $A(z)$ and $B(z)$ instead of $A_0(z)$ and $B_0(z)$. In this way, a simple controller can be developed without considering the plant unmodelled dynamics.

Notice that the factorization of equations (4.9) is not unique. Thus, there are many candidate $A(z)$ and $B(z)$ polynomials to constitute the reduced order model, which may bring about different unmodelled dynamics. It is very important to choose polynomials which describe the simplified model so that $n(t)$ is made small in the useful frequency band. As $A(z)$ and $B(z)$ are obtained from an identification algorithm, in this case the RLS algorithm where the output prediction error is minimized in the least squares sense, an effective reduction of $n(t)$ within the system control bandwidth can be achieved by a filtering procedure.

Data which feed the identifier $y_L(t)$ and $u_L(t)$ are obtained by prefiltering both the plant output and input: $y_L(t) = L(z)y(t)$, $u_L(t) = L(z)u(t)$, where $L(z)$ is a transfer function which represents a lowpass filter which falls at frequencies higher than the band of useful frequencies. In this way, the identification mechanism estimates an input-output model (I/O) of the process $\{y_L(t), u_L(t-d)\}$, (with d being the pure delay of the system), described by the difference equation:

$$A(z)y_L(t) = B(z)u_L(t) + n_L(t)$$

where $n_L(t) := L(z)n(t)$. Notice that if the filtering polynomial is adequately chosen, $n_L(t)$ has its principal components inside the useful frequency band.

If $A(z)$ and $B(z)$ are chosen so that $n_L(t)$ is minimized, the identifier is forced by the prefiltering action $L(z)$ to choose, amongst all possible polynomials, that which satisfactorily approximates $A_0(z)$ and $B_0(z)$ in the useful frequency band. The prefiltering of data with lowpass filters before entering the identifier is very important, due to the fact that the identifier can thus choose different possible reduced order models of the plant. Moreover, by doing this, sudden changes in the variation of the parameters are avoided, allowing for two differentiated time scales. Using adaptive schema without taking into account the above considerations may result in a global behaviour very dependent on the nature of the exogenous signals acting on the system. Nevertheless, with an appropriately selected filtering action, a certain degree of robustness can be obtained in respect both to the excitation (reference signal and disturbances) and the unmodelled dynamics. This is due to the fact that the model is more exact within these frequency bands which are sufficiently excited, and data entering the estimator is made to have dynamical information only in those bands in which it is important to have a good description of the system.

Another important question which arises is that by using CARIMA models, steady state errors and mean values of signals can be eliminated, but the high frequency properties of the process are accentuated (because both the input and the output of the process are being multiplied by Δ). Thus, the differentiation of these signals must be accompanied by the filtering of high frequency components.

The majority of the physical processes possess complex dynamics, and even nonlinear or infinite dimension. A fundamental feature of most adaptive controllers is the use of linear simplified models. The parameters obtained by using a reduced order model are mainly dependent on the properties of the input signal. For instance, a linear model can describe a nonlinear system if the input signal is a square wave of constant amplitude, but large discrepancies can appear for different amplitudes of the input signal. The frequency characteristics of the input signal are also fundamental.

To obtain robust control, it is necessary for the model to approach the actual system as closely as possible to the frequency, at which the phase is -180°. In order to estimate a low order model with this property, it is necessary for the input signal to have sufficient dynamic richness around this frequency. The excitation degree needed in the input signal is a function of the complexity of the estimated model. This implies that the higher the order of the model the more dynamic richness of the input signal is necessary. As the input signal is generated by feedback, there is no guarantee that this signal is being continuously excited. To guarantee a good model, it is necessary to monitorize the excitation of the input signals at the adequate frequencies.

There are two possibilities in the case in which there is a lack of dynamic information at the input signal: either to inject disturbance signals or to stop the adaptation mechanism when this occurs.

4.2 Supervisory levels

The identifier is a fundamental part of the adaptive control mechanism. The identifier should converge to a set of parameters that adequately describes the process dynamics for the operating frequency range. A number of conditions must be fulfilled in order to guarantee this convergence, which depend on the type of disturbances and the dynamic information entering the identifier.

Some improvements can be achieved in the performance of the closed loop controlled system if, parallel to the identification algorithm, a certain supervisory mechanism is implemented to check the evolution of fundamental process parameters. Several methods to characterize distributed solar collector fields have been studied in chapter 2. This characterization, when extended to the whole range of possible operating conditions, provides relevant information not only for controller design purposes, but also for supervising the correct behaviour of the identification mechanism in the case of adaptive control strategies. Particularly, the following supervisory mechanisms have been implemented when controlling the distributed solar collector field:

- **Supervision of the estimated parameters.**

  In the case of simplified Type A and B models, the estimated open loop system pole must be within the range 0.8 to 0.95 and the estimated gain between 0.9 and 1.2 (including the series feedforward controller as a part of the plant). These approximated margins have been chosen by analyzing the dynamic response of the plant under a wide range of operating conditions and are necessary because if a wrong estimation of the parameters occurs, the process may even be made unstable.

- **Supervision of the evolution of the identifier.**

  The value of the trace of the covariance matrix must be checked at each sampling time. This is due to a fact which has been observed in the identification of the system parameters models. Once steady state conditions are reached, the estimated parameters slowly move away from the steady state values, due to an excessive increase (or decrease) of the trace of the covariance matrix. This fact can be justified by noticing that in steady state, while the outlet oil temperature is maintained at a constant value by the control scheme, the reference temperature to the feedforward controller (input signal for the identifier) increases (or decreases) to compensate for the daily cyclic variation of the solar radiation, producing a drift in the estimated parameters. Control of these excursions of the trace of the covariance matrix has been performed by obtaining its maximum and minimum limits in normal operation from experiences. If the value exceeds one of these limits, the covariance matrix is reset and made equal to a prefixed diagonal matrix.

  The main inconvenience presented by this method is that the dynamic behaviour of the system varies with the operating point, and so, as the established limits are fixed, these limits cannot be appropriated at any of these operating points (they can be too conservative or too restrictive). As has been indicated, the reference temperature to the feedforward controller is filtered before entering the identifier to partially cancel unmodelled dynamics. This filter is of the form $t_{fff}(k) = \alpha\, t_{ff}(k) + (1-\alpha)\, t_{fff}(k-1)$ $(0 \leq \alpha \leq 1)$, where t_{ff} is the set point temperature for the feedforward controller calculated by the feedback controller and t_{fff} is the filtered one, which is used as input for the identification mechanism.

In view of the problems mentioned, and although the RLS algorithm with variable forgetting factor accounts for part of these problems, another supervisory mechanism has also been developed, which mainly consists of stopping the identification under the following conditions:

- During the starting phase of the operation (outlet oil temperature below the first set point) in which the inlet oil temperature and the solar radiation suffer great changes that cannot be completely compensated for by the feedforward controlled and may originate the identification of dynamics not corresponding to those modelled by the simplified linear models. The variation in the inlet oil temperature is due to the fact that the oil flowing inside the tubes is cold and, during this phase the oil is recirculated till its temperature reaches a determined value for entering the top of the storage tank, thus experimenting great changes.
- When a considerable decrease in the solar radiation value occurs due to passing clouds. In this case, the values of the identified parameters have to be fixed to the values estimated before the detection of the disturbance (notice that a fixed parameter controller designed to operate around the actual set point is being used). The reason is the same as that given in the previous paragraph: a large cloud can produce quasi-instantaneous drops in the outlet temperature which are not due to changes in the input variable to the identifier.
- When the input and output signal do not contain enough dynamic information. The basic idea is to identify only at those moments in which input signal changes are produced (changes in the reference temperature to the feedforward controller). This can be expressed as:

$$\mid \Delta u \mid \geq A$$

$$\sum_{k=-N}^{k=0} \mid \Delta u(k) \mid \geq B$$

where u is the reference temperature to the feedforward controller. If one of these conditions is true, the identifier is activated. Otherwise, the last set of identified parameters is used. The typical chosen values depend on the plant characteristics. In this case, the chosen values of A, B and N taking into account the field dynamics are: $A = 9$, $7 \leq B \leq 9$ and $N = 5$.

Notice that these values can change within a moderate range due to the stability and good convergence characteristics of the RLS identifier. The main advantage presented by this method is that it is valid for the whole range of operating conditions, due to the fact that it only analyzes the existence of dynamic information entering the identifier, with no regards to the operating point (even thought, implicitly, under certain operating conditions more dynamic information can be given to the identifier than in others). Another advantage is that in this way, the drift of the estimated parameters can be avoided, due to the fact that their values are fixed in those cases which produce this effect.

4.3 Adaptive Ziegler-Nichols rule based PID controllers

As an initial development of adaptive controllers, the Ziegler-Nichols [128] rules have been used. The PID controller transfer function is the same as that explained in section 3.2. In the adaptive version of the algorithm, two different adaptation mechanisms have been used:

1. In the first approach, both the integral and derivative constants are fixed (that is, the time delay is considered to be fixed), and only a proportional gain adaptation is performed from the estimation of the pole and gain of the system linear model (Type A model):

$$G(z) = z^{-1}\frac{bz^{-1}}{(1 - az^{-1})}$$

   In the implementation, the gain obtained by the rules is multiplied by a scale factor to avoid oscillatory responses.
2. In the second approach, integral and derivative constants are also estimated by adding a zero to the transfer function of the system (Type B model) in order to estimate a variable delay by using a kind of first order Pade approximation to a delay. The model used is:

$$G(z) = z^{-2}\frac{(b_0 + b_1 z^{-1})}{(1 - az^{-1})}$$

The complete control scheme is the one presented in Figure 4.2, which is described in the next section.

In the first case (Type A Model), parameters a and b are estimated on line by the RLS algorithm. With these parameters, both:

- the analog time constant $\tau = -T/log(a)$, where T is the sampling time and log the natural logarithm
- and gain $K = b/(1 - a)$

The previous development is valid for both Type A and Type B models (the Type A model being a particularization of the Type B model by making $b_0 = 0$ or $b_1 = 0$). Results achieved with these controllers using both Type A and B models are described in the following sections.

4.5 Simulation analysis of PID controllers

In order to show the behaviour of the adaptive control schema described in the previous section, the results of simulation tests are described next. As an example, Figure 4.3 shows a simulation performed using data from the reference test (section 3.3), for the basic adaptive PID controllers previously described. Figure 4.3 shows the evolution of the outlet oil temperature, the oil flow and the estimated parameters. This simulation serves to draw several logical conclusions about the behaviour which can be expected from any adaptive control scheme using a reduced linear model of the system:

- When the oil flow increases, the delay decreases and the response speed increases. Thus, when solar radiation I is high (usually corresponding to a high flow), the delay is reduced, and also the effective time constant associated to the plant model. This corresponds to a small value of the estimated plant pole a.
- The parameter variations show that the expected trends in the values are obtained.

At the highest flow condition (the central part of the operation), the estimated system pole decreases, corresponding to a faster system. In the operation around 255°C, less flow is needed to achieve a higher outlet temperature and the evolution of this variable follows the daily solar cycle. The gain is mainly dependent on the relation between the radiation level and the flow. Theoretically, the existence of the feedforward controller in series with the plant will provide a system with near unitary gain as, in fact, happens. Changes around this value are due to the fact that the adjustment of the feedforward controller is not perfect. With the oil flow near 10 l/s, the gain is slightly lesser than that corresponding to medium flows, because of the value of the ratio *effective radiation/flow*.

As can be observed, at the first two set point changes, the output temperature follows the set point, but a peak appears in the response after which the output follows the previous evolution. In most control texts, this *kick-back* effect is attributed to a wrong estimation of the gain. However, it has been observed that this effect may not only be due to this fact but also to the existence of resonance and antiresonance characteristics of the controlled plant.

This effect is manifested when a fast response is required of the controlled system. In this case, the first antiresonance mode is excited, producing peaks and oscillations in the time response.

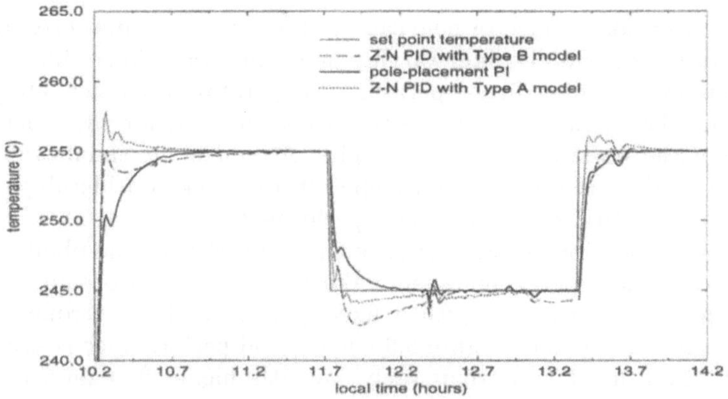

set point and outlet oil temperature (^oC)

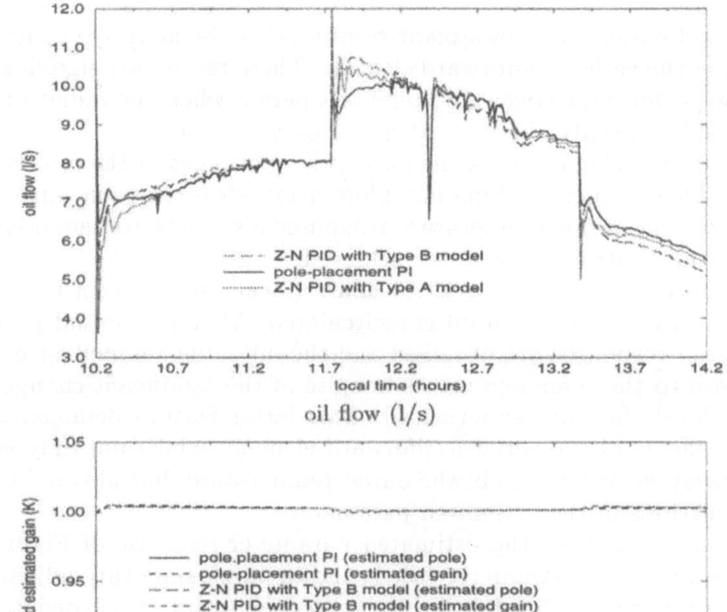

oil flow (l/s)

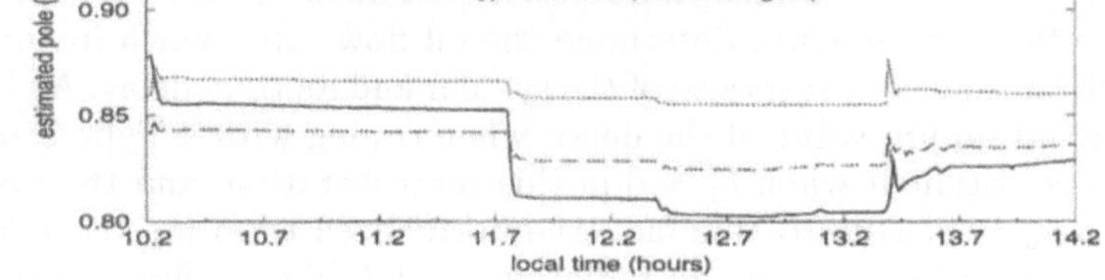

evolution of the estimated pole a and estimated gain K

Fig. 4.3. Simulation with adaptive PID controllers

4.6 Plant results with adaptive PI controllers

The distributed solar collector field can be used for various purposes, in different operation modes and under varying operating conditions (different solar radiation patterns, inlet oil temperature, temperature reference, etc.). All of these factors have, therefore, to be taken into account for assessing the capabilities of the possible control strategies. Although the simulation studies mentioned in the previous section helped in this task considerably, a batch of tests on the actual plant have been performed.

As inferred previously, the dynamic characteristics of the plant vary significantly over the operating range making it difficult to obtain consistent overall control performance with a fixed parameter PI controller. This is shown by the responses of Figure 3.9 where good performance is achieved at a low flow level of 7 l/s approximately, but at a higher flow level of over 10 l/s, the responses are significantly more oscillatory and have a much longer settling time.

Figures 4.4 and 4.5 shows plant results using the adaptive controller incorporating the series feedforward element. These results are significant since they show performance over a 3-hours test period when the radiation level is changing substantially, due to scattered cloud conditions.

Injected disturbances take the form of step changes in the reference temperature. These generate dynamic information for estimation purposes and the resulting oil outlet temperature responses also serve to demonstrate the quality of the control performance achieved.

The period from 11:39 to 12:46 hours (local time) covers the starting-up phase of the plant, when oil is recirculated. After this initial period well damped step responses are obtained and the oil outlet temperature is maintained close to the reference values in spite of the significant changes in the radiation level that are experienced. This latter feature demonstrates the beneficial effects of the serial feedforward element, which not only serves to reduce unwanted variations in the outlet temperature, but also reduces erroneous excursions in the estimated parameters.

This can be seen in the estimated parameter response of Figure 4.5 in which the estimated parameter (a), is shown to be smoothly adjusted during the test period. This follows the expected variation needed to match the changing system response characteristics. The levels of radiation and demanded outlet temperature determine the oil flow rate, which in turn influences both the speed of response of the system and its time delay. As has been stated, the minimum value of the delay, when coping with a Type B model of the plant, is obtained when $b_1 = 0$ in this representation, and the maximum value with $b_0 = 0$. Due to this fact, the normalized term $|b_0/(b_0 + b_1)|$, will increase with an increase in the radiation level I (the oil flow increases and the delay time decreases). The response obtained at the plant can be seen to follow the variation at I, and to give an indication of how the model adapts to the variation in the time delay.

Furthermore, for the conditions of the plant response tests of Figures 4.4 and 4.5, the solar radiation conditions were applied to the simulation model incorporating the fixed PI controller. Again it can be seen that the fixed PI controller, which is tuned for medium flow conditions, tends to oscillate for the low level conditions obtained at the end of the test when a set point with a higher temperature (270 °C) coincides with a low radiation level (around 600 W/m^2). The adaptive control scheme is shown to cope much better than the fixed parameter controller with the changes in plant dynamic characteristics which take place throughout a working day despite the high variations of the radiation level.

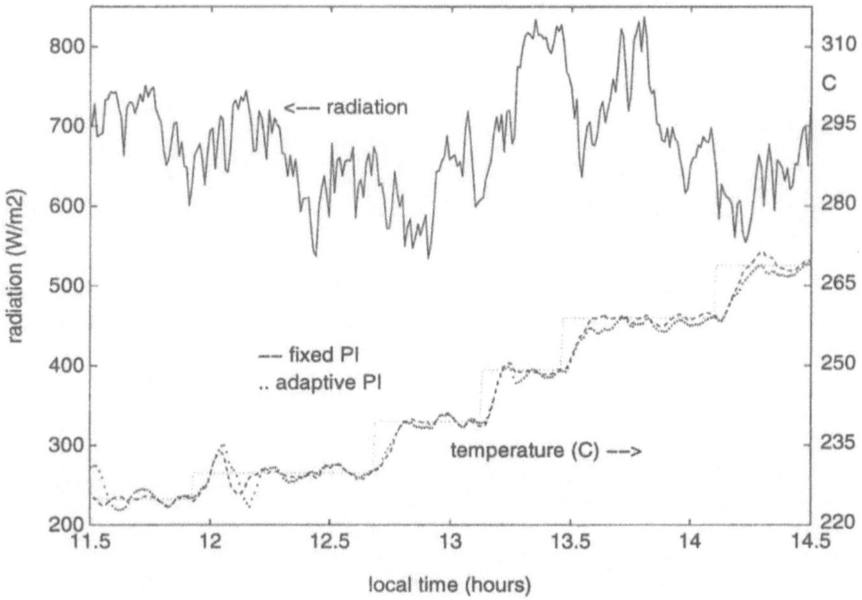

Fig. 4.4. Test with the adaptive PI controller (04/08/91)

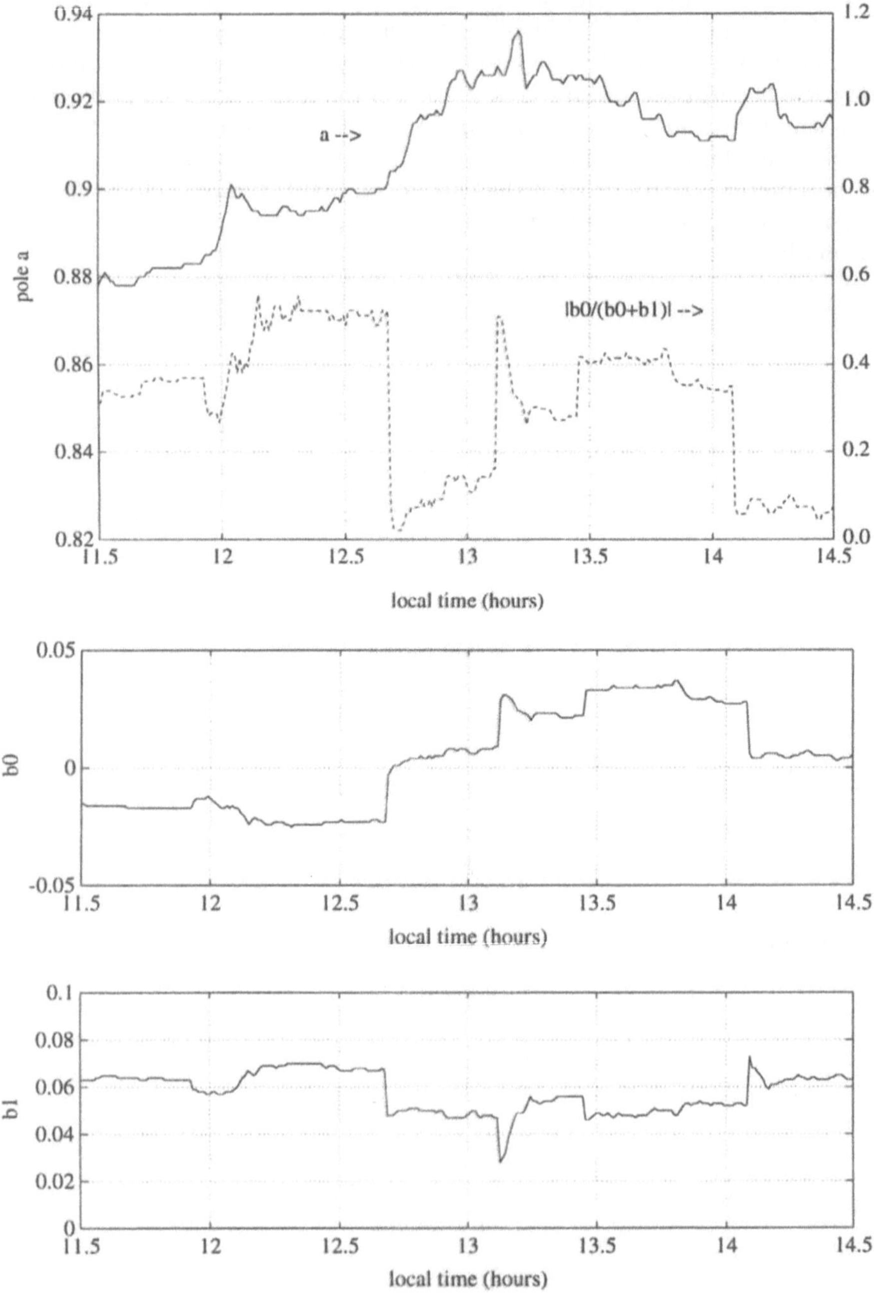

Fig. 4.5. Test with the adaptive PI controller (04/08/91)

CHAPTER 5

MODEL-BASED PREDICTIVE CONTROL STRATEGIES

In this chapter, four predictive control strategies are introduced. The first control strategy presented uses a reduced order linear model of the plant to obtain the control law, which parameters are changed on-line within an indirect adaptive control strategy. The parameters of the simplified linear model are identified and used to properly modify the controller parameters. Problems associated with the convergence of the identifier and errors due to unmodelled dynamics are studied. These modelling errors can be critical in the kind of systems studied here, due to the presence of antiresonance modes, which can be situated within the bandwidth of the controlled system if a fast response or good disturbance rejection properties are required to the controlled system.

The second control strategy is also an adaptive GPC one, but it incorporates in the design explicitly errors due to unmodelled dynamics. The errors due to unmodelled dynamics are included in the design as bounded uncertainties, augmenting the robustness of a basic scheme of adaptive control.

Other possibility of coping with the presence of antiresonance modes is to design the controller using several linear high order models into a gain scheduling framework. A fixed GPC controller is obtained each of the high order linear models obtained to cover the whole range of operating conditions and an interpolation is performed for the controller parameters taking into account the operating point given by a filtered oil flow. This scheme has produced very good results in the real operation, showing a great robustness against disturbances and extreme operating conditions, even in cases with intermittent connections to the desalination plant.

Finally, the chapter finished with a nonlinear GPC strategy. This scheme uses a nonlinear model of the plant to predict the system free response over the prediction horizon and a low order linear model is used to obtain the forced response. A clear day prediction model of the direct solar radiation is also included to perform the prediction of the future disturbances acting on the system.

5.1 Generalized predictive control (GPC)

Most single input – single output (SISO) plants, when considering operation around a particular set point and after linearization, can be described by

$$A(z^{-1})y(t) = z^{-d}B(z^{-1})u(t-1) + C(z^{-1})\epsilon(t) \tag{5.1}$$

where $u(t)$ and $y(t)$ are the control and output sequence of the plant, d is the dead time of the system and $\epsilon(t)$ is a zero mean white noise. A, B and C are the following polynomials in the backward shift operator z^{-1}:

$$\begin{aligned} A(z^{-1}) &= 1 + a_1 z^{-1} + a_2 z^{-2} + \ldots + a_{na} z^{-na} \\ B(z^{-1}) &= b_0 + b_1 z^{-1} + b_2 z^{-2} + \ldots + b_{nb} z^{-nb} \\ C(z^{-1}) &= 1 + c_1 z^{-1} + a_2 z^{-2} + \ldots + c_{nc} z^{-nc} \end{aligned}$$

This model is known as a CARMA model (Controller Auto Regressive Moving Average). It has been argued [35] that for many industrial applications in which disturbances are non-stationary, an integrated CARMA (CARIMA) model is more appropriate. A CARIMA model is given by:

$$A(z^{-1})y(t) = z^{-d}B(z^{-1})u(t-1) + C(z^{-1})\frac{\epsilon(t)}{\Delta} \tag{5.2}$$

$$\Delta = 1 - z^{-1}$$

For simplicity, in the following the $C(z^{-1})$ polynomial is chosen to be 1. Notice that if $C^{-1}(z^{-1})$ can be truncated it can be absorbed into $A(z^{-1})$ and $B(z^{-1})$. The general case of a colored noise is treated in [36].

The Generalized Predictive Control (GPC) algorithm consists of applying a control sequence that minimizes a multistage cost function of the form:

$$J(N_1, N_2, N_u) = E\{\sum_{j=N_1}^{N_2} \delta(j)[\hat{y}(t+j \mid t) - w(t+j)]^2 + \sum_{j=1}^{N_u} \lambda(j)[\Delta u(t+j-1)]^2\} \tag{5.3}$$

where $E\{\cdot\}$ is the expectation operator and $\hat{y}(t+j \mid t)$ is an optimum j-step ahead prediction of the system output on data up to time t; N_1 and N_2 are the minimum and maximum costing (prediction) horizons, N_u is the control horizon, $\delta(j)$ and $\lambda(j)$ are weighting sequences and $w(t+j)$ is a future set point or reference sequence. The expectation operator is used to indicate that the chosen control signal values are calculated with data available at time t and assuming a model of stochastic noise. Although in most cases $w(t+j)$ will be constant and equal to the current reference $w(t)$, in some applications, future set points will be known.

The objective of predictive control is to compute the future control sequence $u(t)$, $u(t+1)$,... in such a way that the optimal j-step ahead prediction

$\hat{y}(t+j \mid t)$ is driven close to $w(t+j)$. This is accomplished by minimizing $J(N_1, N_2, N_u)$.

In order to optimize the cost function, the optimal prediction of $y(t+j)$ (that is, supposing that the predicted output coincides with the real one) for $j \geq N_1$ and $j \leq N_2$ will be obtained. Consider the following Diophantine equation:

$$1 = E_j(z^{-1}) \triangle A(z^{-1}) + z^{-j} F_j(z^{-1}) \tag{5.4}$$

The polynomials E_j and F_j are uniquely defined with degrees $j-1$ and na respectively. They can be obtained by dividing 1 by $\triangle A(z^{-1})$ until the remainder can be factorized as $z^j F_j(z^{-1})$. The quotient of the division is the polynomial $E_j(z^{-1})$. If equation (5.1) is multiplied by $\triangle E_j(z^{-1})\ z^j$

$$\triangle A(z^{-1}) E_j(z^{-1}) y(t+j) = E_j(z^{-1}) B(z^{-1}) \triangle u(t+j-d-1) + E_j(z^{-1}) \epsilon(t+j) \tag{5.5}$$

Considering (5.4), equation (5.5) can be written as:

$$(1 - z^{-j} F_j(z^{-1})) y(t+j) = E_j(z^{-1}) B(z^{-1}) \triangle u(t+j-d-1) + E_j(z^{-1}) \epsilon(t+j) \tag{5.6}$$

which can be rewritten as:

$$y(t+j) = F_j(z^{-1}) y(t) + E_j(z^{-1}) B(z^{-1}) \triangle u(t+j-d-1) + E_j(z^{-1}) \epsilon(t+j) \tag{5.7}$$

As the degree of polynomial $E_j(z^{-1})$ is $j-1$ the noise terms in equation (5.7) are all in the future. The best prediction of $y(t+j)$ is therefore:

$$\hat{y}(t+j \mid t) = G_j(z^{-1}) \triangle u(t+j-d-1) + F_j(z^{-1}) y(t) \tag{5.8}$$

where $G_j(z^{-1}) = E_j(z^{-1}) B(z^{-1})$.

It is very simple to show that the polynomials $E_j(z^{-1})$ and $F_j(z^{-1})$ can be obtained recursively. Consider that these polynomials have been obtained dividing 1 by $\triangle A(z^{-1})$ until the remainder of the division can be factorized as $z^{-j} F_j(z^{-1})$, with:

$$\begin{aligned} F_j(z^{-1}) &= f_{j,0} + f_{j,1} z^{-1} + \cdots + f_{j,na} z^{-na} \\ E_j(z^{-1}) &= e_{j,0} + e_{j,1} z^{-1} + \cdots + e_{j,j-1} z^{-(j-1)} \end{aligned}$$

Suppose that the same procedure is used to obtain $E_{j+1}(z^{-1})$ and $F_{j+1}(z^{-1})$, that is, dividing 1 by $\triangle A(z^{-1})$ until the remainder of the division can be factorized as $z^{-(j+1)} F_{j+1}(z^{-1})$, with:

$$\begin{aligned} F_{j+1}(z^{-1}) &= f_{j+1,0} + f_{j+1,1} z^{-1} + \cdots + f_{j+1,na} z^{-na} \\ E_{j+1}(z^{-1}) &= e_{j+1,0} + e_{j+1,1} z^{-1} + \cdots + e_{j+1,j} z^{-j} \end{aligned}$$

It is clear that only another step of the division performed to obtain the polynomials $E_j(z^{-1})$ and $F_j(z^{-1})$ has to be taken in order to obtain the polynomials $E_{j+1}(z^{-1})$ and $F_{j+1}(z^{-1})$. The polynomial $E_{j+1}(z^{-1})$ will be given by:

$$E_{j+1}(z^{-1}) = E_j(z^{-1}) + e_{j+1,j}z^{-j} \tag{5.9}$$

with $e_{j+1,j} = f_{j,0}$. The coefficients of polynomial $F_{j+1}(z^{-1})$ can be obtained as:

$$f_{j+1,i} = f_{j,i+1} - f_{j,0}\, a'_{i+1} \quad \text{for } i = 0 \cdots na - 1$$

where a'_{i+1} is the $i+1$ coefficient of $\Delta A(z^{-1})$.

The polynomial $G_{j+1}(z^{-1})$ can be obtained recursively as follows:

$$\begin{aligned} G_{j+1}(z^{-1}) &= E_{j+1}(z^{-1})B(z^{-1}) = (E_j(z^{-1}) + f_{j,0}z^{-j})B(z^{-1}) \\ G_{j+1}(z^{-1}) &= G_j(z^{-1}) + f_{j,0}z^{-j}B(z^{-1}) \end{aligned}$$

That is, the first j coefficient of $G_{j+1}(z^{-1})$ will be identical to those of $G_j(z^{-1})$ and the remaining coefficients will be given by:

$$g_{j+1,j+i} = g_{j,j+i} + f_{j,0}\, b_i \quad \text{for } \; i = 0 \cdots nb$$

To solve the GPC problem the set of control signals $u(t)$, $u(t+1)$, ..., $u(t+N)$ has to be obtained in order to optimize equation (5.3). As the system considered has a dead time of d sampling periods, the output of the system will be influenced by signal $u(t)$ after sampling period $d+1$. The values N_1, N_2 and N_u defining the control horizon can be defined by $N_1 = d+1$, $N_2 = d+N$ and $N_u = N$. Notice that there is no point in making $N_1 < d+1$ because terms added to expression (5.3) will only depend on the past control signals. Furthermore, by making $N_1 > d+1$, the first points in the reference sequence, which are those that can be guessed with greater certainty, will not be taken into account.

Consider now the following set of j ahead optimal predictions:

$$\begin{aligned} \hat{y}(t+d+1 \mid t) &= G_{d+1}(z^{-1})\,\Delta\, u(t) + F_{d+1}(z^{-1})y(t) \\ \hat{y}(t+d+2 \mid t) &= G_{d+2}(z^{-1})\,\Delta\, u(t+1) + F_{d+2}(z^{-1})y(t) \\ &\vdots \\ \hat{y}(t+d+N \mid t) &= G_{d+N}(z^{-1})\,\Delta\, u(t+N-1) + F_{d+N}(z^{-1})y(t) \end{aligned}$$

which can be written as:

$$\mathbf{y} = \mathbf{G}\mathbf{u} + \mathbf{F}(z^{-1})y(t) + \mathbf{G}'(z^{-1})\,\Delta\, u(t-1) \tag{5.10}$$

Where

$$\mathbf{y} = \begin{bmatrix} \hat{y}(t+d+1 \mid t) \\ \hat{y}(t+d+2 \mid t) \\ \vdots \\ \hat{y}(t+d+N \mid t) \end{bmatrix} \qquad \mathbf{u} = \begin{bmatrix} \Delta u(t) \\ \Delta u(t+1) \\ \vdots \\ \Delta u(t+N-1) \end{bmatrix} \tag{5.11}$$

$$
\begin{aligned}
\mathbf{G} &= \begin{bmatrix} g_0 & 0 & \cdots & 0 \\ g_1 & g_0 & \cdots & 0 \\ \vdots & \vdots & \vdots & \vdots \\ g_{N-1} & g_{N-2} & \cdots & g_0 \end{bmatrix} \\
\mathbf{G}'(z^{-1}) &= \begin{bmatrix} (G_{d+1}(z^{-1}) - g_0)z \\ (G_{d+2}(z^{-1}) - g_0 - g_1 z^{-1})z^2 \\ \vdots \\ (G_{d+N}(z^{-1}) - g_0 - g_1 z^{-1} - \cdots - g_{N-1} z^{-(N-1)})z^N \end{bmatrix} \\
\mathbf{F}(z^{-1}) &= \begin{bmatrix} F_{d+1}(z^{-1}) \\ F_{d+2}(z^{-1}) \\ \vdots \\ F_{d+N}(z^{-1}) \end{bmatrix}
\end{aligned}
\tag{5.12}
$$

Notice that the last two terms in equation (5.10) only depend on the past and can be grouped into $\mathbf{f}$ leading to:

$$\mathbf{y} = \mathbf{G}\mathbf{u} + \mathbf{f} \tag{5.13}$$

Equation (5.3) can be written as:

$$J = (\mathbf{G}\mathbf{u} + \mathbf{f} - \mathbf{w})^T(\mathbf{G}\mathbf{u} + \mathbf{f} - \mathbf{w}) + \lambda \mathbf{u}^T\mathbf{u} \tag{5.14}$$

where:

$$\mathbf{w} = [w(t+d+1 \mid t) \;\; w(t+d+2 \mid t) \cdots \;\; w(t+d+N \mid t)]^T$$

(the components of $\mathbf{w}$ are supposed to be known future references). Equation (5.14) can be written as:

$$J = \frac{1}{2}\mathbf{u}^T\mathbf{H}\mathbf{u} + \mathbf{b}^T\mathbf{u} + f_0 \tag{5.15}$$

$$
\begin{aligned}
\mathbf{H} &= 2(\mathbf{G}^T\mathbf{G} + \lambda\mathbf{I}) \\
\mathbf{b}^T &= 2\,(\mathbf{f} - \mathbf{w})^T\mathbf{G} \\
f_0 &= (\mathbf{f} - \mathbf{w})^T(\mathbf{f} - \mathbf{w})
\end{aligned}
$$

The minimum of J, assuming no constraints on the control signals, can be found by making the gradient of J equal to zero, which leads to: $\mathbf{u} = -\mathbf{H}^{-1}\mathbf{b}$.

Only the first component of vector $\mathbf{u}$ is applied, shifting to the future prediction and control horizons one step and repeating the algorithm again (receding horizon control strategy).

The GPC involves the solution of an unconstrained linear quadratic (LQ) problem with N variables, which can easily be obtained by using any standard method. Although the amount of computation needed is not very high, it can be a drawback for some real time applications. In the case of a fixed controller the main part of the computation has to be carried out once, but in the self-tuning controller case, the computation has to be performed at each sampling period. This problem is aggravated in the case of processes with bounded signals, where a constrained LQ has to be solved. Fast algorithms for self-tuning GPC controllers that can be applied to processes which can be modelled by the reaction curve method have been proposed by some of the authors [23], [21], [24], and the constrained problem has been treated in [17], [24]. Both algorithms are developed in this chapter and applications issues are also expounded.

5.2 Constrained generalized predictive control

In practice all processes are subject to constraints. The actuators have a limited range of influence as well as a limited speed of change, as in the case of the control valves that are limited by a fully closed and fully open position and a maximum slew rate. The values attainable by the process variables, such as tank levels, fluid flow along the pipe and pressure in the deposits, are limited due to constructive or security reasons and also to the possible sensor ranges. The control system usually operates very close to its limits and it is very probable that certain constraints will be violated.

One of the advantages of model predictive control is that constraints can be taken into account. The constraints in the GPC have been studied in [119] where two algorithms are developed for processes with constraints in the input signals of the system. The first algorithm takes into account the case of limitations in the speed of change in the control signals and assumes that the optimum one will be in the constraint violated in the unconstrained optimization phase of the algorithm. The second case can be applied to systems with limits in the range of process input signals and with a control horizon of two sampling periods. A quadratic programming approach (QP) is suggested in [34] to take constraints into account more generally. The article makes the fundamental conclusion that the QP algorithm is not adequate for fast systems, due to its complexity and to the computation load necessary. The same conclusion is reached in [47] in the case of MPC, suggesting the development of appropriate algorithms for solving this problem.

The methods for solving quadratic programming (QP) problems in an exact way and in a finite number of cycles, as is indicated in [65], can be classified in the following groups:

- Those based on the computation of Lagrange multipliers, such as the *active set method* [44].
- Those based on the reduction of a QP problem to a non negative least squares problem (NNLS) [70].
- Those that reduce the problem to a complementary linear one [44], [9], [17].

Dynamic programming can also be used to solve MPC with constraints; this is a more general method allowing for nonlinear models and objective functions not necessarily quadratic to be dealt with, although it involves a heavy computational load which can be reduced by using precalculated tables, as shown in [78].

One of the main problems, other than the great computational burden, is now to obtain a feasible solution. If the constraints only affect the control signal the saturated unconstrained solution can always be used (limiting the achieved performance), but this cannot be done for output constraints. There are various suggestions proposed to deal with this subject (see [109] for example).

Another of the associated problems is that of stability and robustness as shown in [125], [91], [126], etc. Until recently no results about the stability of MPC in the presence of constraints had been obtained. However, new approaches have appeared based on the idea that the cost functions are monotonously decreasing and that if there is a feasible solution, they can be interpreted as Lyapunov functions, in this way guaranteeing stability. On one hand Clarke and Scattolini's results with Constrained Receding Horizon Predictive Control (CRHPC) [37], [89], [90], the idea of which is, basically, the output is forced to coincide with the reference for a certain interval at the end of the prediction horizon. Some of the degrees of freedom of the future control signal are used to force the output to follow the set point, whilst the rest remain available to minimize the cost function in a certain interval.

Another important approach dealing with stability results is that taken by Rawlings and Muske [94] who considered the objective function to be composed of two parts: one with a finite and restricted horizon and the other with an infinite one without constraints. The infinite part can be solved by a *Riccati equation* and the resulting optimal cost function can be obtained as a quadratic function of the state that is later introduced into the finite optimization problem, which is solved by numerical methods (QP).

The method developed by Camacho [17] which reduces the constrained GPC problem to a complementary linear one is used in the book when constraints are taken into account. Some remarks are done here.

Constrained GPC

In practice, the normal way of using a GPC is to compute $u(t)$ as described previously and apply it to the process. If $u(t)$ violates the constraints it is saturated to its limits, either by the control program or by the actuator. The case of $u(t+1), \cdots, u(t+N)$ violating the constraints is not even considered as in most cases these signals are not even computed.

This way of operating does not guarantee that the optimum will be obtained when constraints are violated. The main purpose of the GPC, which is to apply the *best* possible control signal by minimizing expression (5.3), will not be achieved. In order to illustrate this point, consider the cases of constraint violations shown in Figure 5.1 of a GPC problem with a control horizon of two.

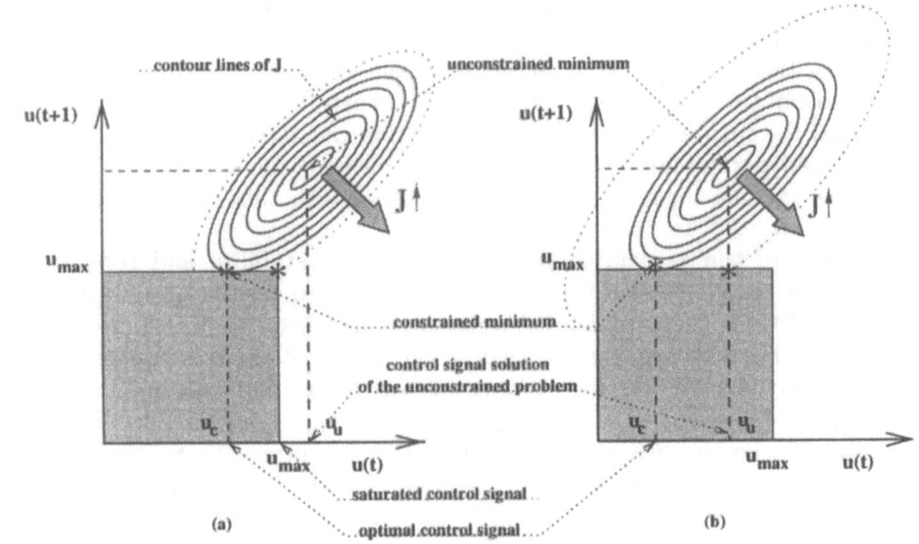

Fig. 5.1. Constrained GPC with N=2

Figure 5.1(a) shows the case where $u(t) > u_{max}$ (the optimal solution without taking into account the constraints u_u exceeds the maximum value limited by these). In this case the normal way of operating is to apply $u(t) = u_{max}$ to the process instead of u_c where the minimum of J is reached when restrictions are considered. In the case shown in Figure 5.1(b), u_u does not violate the constraints, and it will be applied to the system $(u(t) = u_u)$ instead of the optimum signal u_c that should be applied when constraints both in the actual and next control signal are taken into account.

As has been commented, many different constraints can act on a process. The most common constraints acting on a process can be originated by amplitude limits in the: control signals, control signals increments (slew rate limits of the actuators), control signals acceleration, output signals, output signals derivative, output signals acceleration and overshoot. The last one is very important in the process industry and mostly in thermal processes in which cooling processes require an additional control effort.

These constraints can be analytically described by:

$$\begin{array}{rcccl} u_{min} & \leq & u(t) & \leq & u_{max} \quad \forall t \\ du_{min} & \leq & u(t) - u(t-1) & \leq & du_{max} \quad \forall t \\ \bar{y}_{min} & \leq & y(t) & \leq & \underline{y}_{max} \quad \forall t \end{array}$$

with $\bar{y}_{min} = max\{y_{min}, w(t)\}$ and $\underline{y}_{max} = min\{y_{max}, w(t)\}$ (future references are supposed to be known or equal to the actual reference).

These constraints can be expressed as:

$$\begin{array}{ccccc} \mathbf{1}\, u_{min} & \leq & \mathbf{Tu} + u(t-1)\,\mathbf{1} & \leq & \mathbf{1}\, u_{max} \\ \mathbf{1}\, du_{min} & \leq & \mathbf{u} & \leq & \mathbf{1}\, du_{max} \\ \mathbf{1}\, \bar{y}_{min} & \leq & \mathbf{Gu} + \mathbf{f} & \leq & \mathbf{1}\, \underline{y}_{max} \end{array}$$

where $\mathbf{1}$ is an N vector whose entries are ones and $\mathbf{T}$ is an $N \times N$ lower triangular matrix whose entries are ones. If $w(t)$ is not constant along the control horizon, a modified $\mathbf{1}$ vector can be included in the last inequality which components can account for future reference variations with respect to the actual one. If $\mathbf{u} = \mathbf{1}\, du_{min} + \mathbf{x}$ the constraints can be expressed in condensed form as:

$$\begin{array}{c} \mathbf{x} \geq \mathbf{0} \\ \mathbf{R}\,\mathbf{x} \leq \mathbf{c} \end{array} \tag{5.16}$$

with:

$$\mathbf{R} = \begin{bmatrix} \mathbf{I}_{N\times N} \\ \mathbf{T} \\ -\mathbf{T} \\ \mathbf{G} \\ -\mathbf{G} \end{bmatrix} \; ; \; \mathbf{c} = \begin{bmatrix} \mathbf{1}\,(du_{max} - du_{min}) \\ \mathbf{1}\, u_{max} - \mathbf{T}\,\mathbf{1}\, du_{min} - u(t-1)\mathbf{1} \\ -\mathbf{1}\, u_{min} + \mathbf{T}\,\mathbf{1}\, du_{min} + u(t-1)\mathbf{1} \\ \mathbf{1}\, \underline{y}_{max} - \mathbf{f} - \mathbf{G}\,\mathbf{1}\, du_{min} \\ -\mathbf{1}\, \bar{y}_{min} + \mathbf{f} + \mathbf{G}\,\mathbf{1}\, du_{min} \end{bmatrix} \tag{5.17}$$

Equation (5.3) can be rewritten as:

$$J = \frac{1}{2}\mathbf{x}^T\,\mathbf{H}\,\mathbf{x} + \mathbf{a}^T\mathbf{x} + f_1 \tag{5.18}$$

where: $\mathbf{a} = \mathbf{b} + u_{min}\,\mathbf{H}^T\mathbf{1}$ and $f_1 = f_0 + u_{min}^2\,\mathbf{1}^T\,\mathbf{H}\,\mathbf{1} + \mathbf{1}^T\,\mathbf{b}\,u_{min}$

The problem is reduced to minimizing expression (5.18) subject to constraints (5.16). Denoting the Lagragian multiplier vectors of the constraints $\mathbf{x} \geq 0$ and $\mathbf{Rx} \leq \mathbf{c}$ by $\mathbf{v}$ and $\mathbf{v}_1$, respectively, and denoting the vector of slack variables by $\mathbf{v}_2$, the Kuhn-Tucker conditions can be written as

$$\begin{array}{rcl} \mathbf{R}\,\mathbf{x} + \mathbf{v}_2 & = & \mathbf{c} \\ -\mathbf{H}\,\mathbf{x} - \mathbf{R}^T\mathbf{v} + \mathbf{v}_1 & = & \mathbf{a} \\ \mathbf{x}^T\mathbf{v}_1 = 0,\; \mathbf{v}^T\mathbf{v}_2 & = & 0 \\ \mathbf{x}, \mathbf{v}, \mathbf{v}_1, \mathbf{v}_2 & \geq & \mathbf{0} \end{array} \tag{5.19}$$

These expressions can be rewritten as:

$$\begin{bmatrix} \mathbf{I}_{m\times m} & \mathbf{0}_{m\times N} & \mathbf{0}_{m\times m} & \mathbf{R} \\ \mathbf{0}_{N\times m} & \mathbf{I}_{N\times N} & -\mathbf{R}^T & -\mathbf{H} \end{bmatrix} \begin{bmatrix} \mathbf{v}_2 \\ \mathbf{v}_1 \\ \mathbf{v} \\ \mathbf{x} \end{bmatrix} = \begin{bmatrix} \mathbf{c} \\ \mathbf{a} \end{bmatrix} \tag{5.20}$$

$$\begin{aligned} \mathbf{x}^T\mathbf{v}_1 = 0,\ \mathbf{v}^T\mathbf{v}_2 &= 0 \\ \mathbf{x}, \mathbf{v}, \mathbf{v}_1, \mathbf{v}_2 &\geq \mathbf{0} \end{aligned}$$

where m is the number of rows of matrix $\mathbf{R}$. The Kuhn-Tucker conditions can be expressed as a linear complementary problem $\mathbf{s} - \mathbf{M}\mathbf{z} = \mathbf{q}$, $\mathbf{s}^T\mathbf{z} = 0$, $\mathbf{s}, \mathbf{z} \geq 0$ with:

$$\mathbf{M} = \begin{bmatrix} \mathbf{0} & -\mathbf{R} \\ \mathbf{R}^T & \mathbf{H} \end{bmatrix}, \quad \mathbf{q} = \begin{bmatrix} \mathbf{c} \\ \mathbf{a} \end{bmatrix}, \quad \mathbf{s} = \begin{bmatrix} \mathbf{v}_2 \\ \mathbf{v}_1 \end{bmatrix}, \quad \mathbf{z} = \begin{bmatrix} \mathbf{v} \\ \mathbf{x} \end{bmatrix} \tag{5.21}$$

This problem can be solved using Lemke's algorithm (see [17], [24]) by forming the following table:

	$\mathbf{v}_2$	$\mathbf{v}_1$	$\mathbf{v}$	$\mathbf{x}$	z_0	
$\mathbf{v}_2$	$\mathbf{I}_{m\times m}$	$\mathbf{0}_{m\times N}$	$\mathbf{0}_{m\times m}$	$\mathbf{R}$	$-\mathbf{1}$	$\mathbf{c}$
$\mathbf{v}_1$	$\mathbf{0}_{N\times m}$	$\mathbf{I}_{N\times N}$	$-\mathbf{R}^T$	$-\mathbf{H}$	$-\mathbf{1}$	$\mathbf{a}$

Although the algorithm will converge to the optimum solution in a finite number of steps as matrix $\mathbf{H}$ is positive definite [9], it needs a substantial amount of computation. The fact that the $\mathbf{x}$ variables in the starting solution of Lemke's algorithm are not part of the basis is one of the reason for this. That is, the algorithm starts from the solution $\mathbf{x} = 0$ which may be far away from the optimum solution. The efficiency of the algorithm can be increased [17], [24] by finding a better starting point.

If equation (5.20) is multiplied by: $\begin{bmatrix} \mathbf{I}_{m\times m} & \mathbf{R}\,\mathbf{H}^{-1} \\ \mathbf{0}_{N\times m} & -\mathbf{H}^{-1} \end{bmatrix}$, it can be obtained,

$$\begin{bmatrix} \mathbf{I}_{m\times m} & \mathbf{R}\mathbf{H}^{-1} & \mathbf{R}\mathbf{H}^{-1}\mathbf{R}^T & \mathbf{0}_{m\times N} \\ \mathbf{0}_{N\times m} & -\mathbf{H}^{-1} & \mathbf{H}^{-1}\mathbf{R}^T & \mathbf{I}_{N\times N} \end{bmatrix} \begin{bmatrix} \mathbf{v}_2 \\ \mathbf{v}_1 \\ \mathbf{v} \\ \mathbf{x} \end{bmatrix} = \begin{bmatrix} \mathbf{c} + \mathbf{R}\mathbf{H}^{-1}\mathbf{a} \\ -\mathbf{H}^{-1}\mathbf{a} \end{bmatrix} \tag{5.22}$$

The vector on the right hand side of equation (5.22) corresponds to the vector of slack variables for the unconstrained solution and to the unconstrained solution respectively. Furthermore, equation (5.22) shows that if Lemke's algorithm is started from this point, all the $\mathbf{x}$ variables are in the basis. In most cases only a few constraints will be violated for the unconstrained solution of the GPC problem. Thus, the constrained solution will be close to the initial condition and the number of iterations required should decrease.

The proposed algorithm can be described as follows,

1. Compute the unconstrained solution $\mathbf{x}_{min} = -\mathbf{H}^{-1}\mathbf{a}$.
2. Compute $\mathbf{v}_{2min} = \mathbf{c} - \mathbf{R}\mathbf{x}_{min}$. If $\mathbf{x}_{min}$ and $\mathbf{v}_{2min}$ are non negative then stop with $u(t) = u(t-1) + du_{min} + x_1$. Otherwise, continue with point 3.
3. Start Lemke's algorithm with $\mathbf{x}$ and $\mathbf{v}_2$ in the basis with the following table:

	$\mathbf{v}_2$	$\mathbf{x}$	$\mathbf{v}$	$\mathbf{v}_1$	z_0	
$\mathbf{v}_2$	$\mathbf{I}_{m\times m}$	$\mathbf{0}_{m\times N}$	$\mathbf{R}\mathbf{H}^{-1}\mathbf{R}^T$	$\mathbf{R}\mathbf{H}^{-1}$	$-\mathbf{1}$	$\mathbf{v}_{2min}$
$\mathbf{x}$	$\mathbf{0}_{N\times m}$	$\mathbf{I}_{N\times N}$	$\mathbf{H}^{-1}\mathbf{R}^T$	$-\mathbf{H}^{-1}$	$-\mathbf{1}$	$\mathbf{x}_{min}$

4. If x_1 is not in the first column of the table make it zero otherwise give it the corresponding value.
5. $u(t) = u(t-1) + du_{min} + x_1$.

An algorithm for constraint reduction can be found in [17], [24] where, as the computational effort depends substantially on the number of constraints that the algorithm takes into account, an algorithm to maintain only those constraints which limit the feasibility region in which the solution can lie is proposed. The efficiency of the algorithm increases as constraints are eliminated.

5.3 Adaptive generalized predictive control

5.3.1 Introduction

One of the main reasons of the success of traditional PID controllers in the industry is the easiness of their implementation and sintonization using in most cases certain heuristic rules, being Ziegler-Nichols rules the most commonly used in control engineering practice. The development of suitable controllers to be implemented in the process industry is becoming an important motivation in a great amount of research works [24]. In this section an adaptive generalized predictive controller is presented. The adaptive controller has been developed with a method that makes use of the fact that a generalized predictive controller using a quadratic function results in a linear control law that can be described by a few parameters. These parameters can be computed over the range of interest of the process parameters [21], [18]. A Ziegler-Nichols-type function which relates the generalized predictive controller parameters to the process parameters is used to obtain an approximation of the real controller parameters. The method avoids the heavy computation requirement of this type of controller. The method can be applied to processes that can be modelled by the reaction curve method, that is, a wide range of processes in industry.

5.3.2 Application to the distributed solar collector field

According to the method described in [24] and the requirements for self-tuning control a CARIMA model is used to model the random disturbances in the system,

$$A(z^{-1})y(t) = B(z^{-1})u(t-1)z^{-d} + C(z^{-1})\frac{\epsilon(t)}{\Delta} \tag{5.23}$$

where $u(t)$ and $y(t)$ are the control and output sequence of the plant, d is the dead time of the system and $\epsilon(t)$ is a zero mean white noise. If a CARIMA model based on the step response characterization is used, and particularly a first order model (Type A Model), polynomials $A(z^{-1}), B(z^{-1})$ and $C(z^{-1})$ are given by:

$$A(z^{-1}) = 1 - az^{-1} \ ; \quad B(z^{-1}) = b \ ; \quad C(z^{-1}) = 1$$

and in this case d=1. This scheme allows integral action to eliminate the offset in the output signal. The model expression is given by:

$$(1 - az^{-1})y(t) = bz^{-1}u(t-1) + \frac{\epsilon(t)}{\Delta} \tag{5.24}$$

This expression can be easily transformed into:

$$y(t+1) = (1+a)y(t) - ay(t-1) + b\,\Delta\, u(t-1) + \epsilon(t+1) \tag{5.25}$$

For simplicity, fixed values of the weighting sequences have been chosen ($\lambda(j) = \lambda$, $j = 1 ... N_2 - d$ and $\delta(j) = 1, j = N_1 ... N_2$). The most adequate value for the control horizon ($N = 15$) has been calculated taking into account the values of the fundamental time constant and the sampling period used for control purposes. There is no reason to choose greater values of N because steady state would be over weighted. From the robustness viewpoint, when the control horizon N is sufficiently large [67], [48], the closed loop system has some degree of robustness with respect to gain and additive perturbations in the control law. As the system considered has a dead time of $d = 1$ sampling periods, values of $N_1 = 2$, $N_2 = 16$ and $N_u = 15$ are chosen (notice that a lesser value of N_u could be chosen, but in this case, the value of the chosen sampling time allows to use high control horizons). The range of possible values of λ has been obtained via simulation ($3 \leq \lambda \leq 7$). If λ diminish faster controllers are obtained. Expression (5.25) can be applied to obtain the expected value of $y(t + j + 1)$:

$$\hat{y}(t+j+1 \mid t) = (1+a)\hat{y}(t+j \mid t) - a\hat{y}(t+j-1 \mid t) + b \triangle u(t+j-1) \quad (5.26)$$

If equation (5.26) is applied recursively for $j = 1, 2, \cdots, i$ we get

$$\hat{y}(t+i+1 \mid t) = G_i(z^{-1})\hat{y}(t+1 \mid t) + D_i(z^{-1}) \triangle u(t+i-1) \quad (5.27)$$

where $G_i(z^{-1})$ is of degree 1 and $D_i(z^{-1})$ is of degree $i - 1$.

If $\hat{y}(t+i+1 \mid t)$ is introduced in equation (5.3), J is a function of $\hat{y}(t+1 \mid t)$, $y(t)$, $\Delta u(t+14)$, $\Delta u(t+13)$ ... $\Delta u(t)$ and the reference sequence.

Minimizing J with respect to $\Delta u(t)$, $\Delta u(t+1)$ ... $\Delta u(t+14)$ leads to

$$\mathbf{M} \begin{bmatrix} \Delta u(t) \\ \Delta u(t+1) \\ \vdots \\ \Delta u(t+14) \end{bmatrix} = \mathbf{P} \begin{bmatrix} \hat{y}(t+1 \mid t) \\ y(t) \end{bmatrix} + \mathbf{R} \begin{bmatrix} w(t+2) \\ w(t+3) \\ \vdots \\ w(t+16) \end{bmatrix} \quad (5.28)$$

Where $\mathbf{M}$ and $\mathbf{R}$ are of dimension 15×15 and $\mathbf{P}$ of dimension 15×2.

Let us call $\mathbf{q}$ the first row of matrix $\mathbf{M}^{-1}$. Then $\Delta u(t)$ is given by

$$\triangle u(t) = \mathbf{qP} \begin{bmatrix} \hat{y}(t+1 \mid t) \\ y(t) \end{bmatrix} + \mathbf{qR} \begin{bmatrix} w(t+2) \\ w(t+3) \\ \vdots \\ w(t+16) \end{bmatrix} \quad (5.29)$$

If the future references, $w(t+j)$, are unknown and are considered to be equal to the current reference $r(t)$, the control increment $\Delta u(t)$ can be written as:

$$\triangle u(t) = l_1\hat{y}(t+1 \mid t) + l_2 y(t) + l_3 r(t) \quad (5.30)$$

Where $\mathbf{qP} = [l_1 l_2]$ and $l_3 = \sum_{i=1}^{15} q_i \sum_{j=1}^{15} r_{ij}$.

Notice that if future references were known, $l_3 = \mathbf{qR}$ would be a gain vector multiplying future references. The coefficients l_1, l_2, l_3, are functions of a, b, $\delta(i)$ and $\lambda(i)$. If the GPC is designed considering the plant to have a unit static gain, the coefficients in (5.30) will only depend on $\delta(i)$ and $\lambda(i)$ (which are supposed to be fixed) and on the pole of the plant which will change for the adaptive control case. Notice that by doing this, a normalized weighting factor λ is used which should be corrected accordingly for systems with different static gains. Most of all, for systems with a static gain different from unity, the GPC could be obtained in this way by changing coefficients in equation (5.30) accordingly.

The value $\hat{y}(t+1 \mid t)$ is obtained by the use of the predictor described before (5.26). The control scheme proposed is shown in Figure 5.2. The plant estimated parameters $(\hat{a}, \hat{b})$ are used to compute the controller coefficients (l_1, l_2, l_3) via the adaptation mechanism. Notice that in this scheme, the feedforward term is considered as a part of the plant (the control signal is the set point temperature for the feedforward controller instead of the oil flow). This signal is saturated and filtered before its use in the estimation algorithm. The controller also has a saturation to limit the increase of the error signal. The control signal is divided by the estimated gain because the design has been performed supposing a system with unitary static gain.

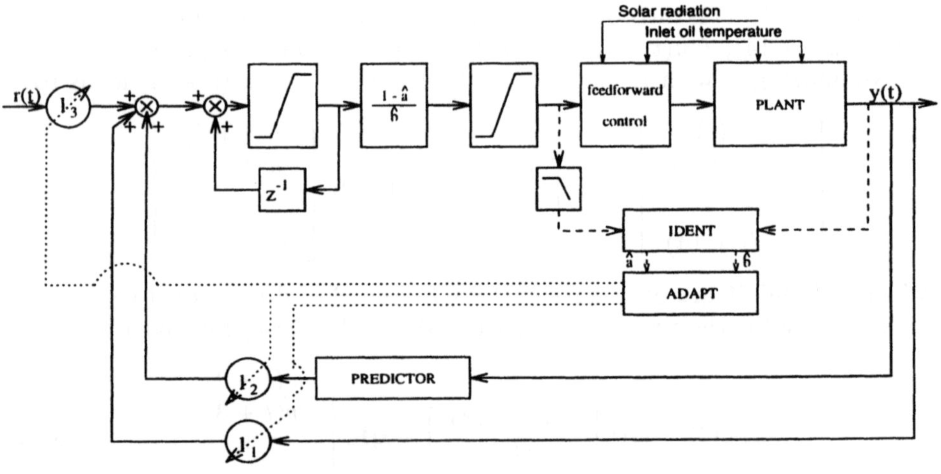

Fig. 5.2. Adaptive GPC control scheme

The standard way of computing the controller coefficients would be by computing matrices $\mathbf{M}, \mathbf{P}$ and $\mathbf{R}$ and solving equation (5.28) followed by the generation of the control law of equation (5.30). This involves the triangularization of an $N \times N$ matrix, which is prohibitive for a real time application.

As suggested in [21], [23], the controller coefficients can be obtained by interpolating in a set of previously computed values. The number of points of the set used depends on the variability of the process parameters and on the accuracy needed. The set does not need to be uniform and more points can be computed in regions where the controller parameters vary substantially in order to obtain a better approximation or to reduce the computer memory needed.

A set of GPC parameters (l_1, l_2, l_3) were obtained for $\delta(i) = 1$, $\lambda(i) = 5$ and $N = 15$. The pole of the system has been changed with a 0.0005 step from 0.8 to 0.95, which are values that guarantee the system stability if the parameter set estimation is not accurate enough (the same procedure has been performed for different values of λ between 3 and 7). Notice that owing to the fact that the closed loop static gain must equal the value unity, the sum of the three parameters equals zero. This result implies that only two of the three parameters need to be known.

The curves shown in Figure 5.3 correspond to the controller parameters l_1,l_2, l_3 for the values of the pole mentioned above. A set of simple Ziegler-Nichols-type functions which approximate the computed values of l_1, l_2 and l_3 has been obtained as follows.

By looking at Figure 5.3, it can be seen that the functions relating the controller parameters to the process pole can be approximated by functions of the form:

$$l_i = k_{1i} + k_{2i} \frac{a}{(k_{3i} - a)} \quad , \quad i = 1, 2 \tag{5.31}$$

The coefficients k_{ji} can be calculated by a least squares adjustment using the set of known values of l_i for different values of a. Equation (5.31) can be written as

$$\begin{bmatrix} a^1 l_i^1 \\ a^2 l_i^2 \\ \vdots \\ a^{N_p} l_i^{N_p} \end{bmatrix} = \begin{bmatrix} l_i^1 & 1 & a^1 \\ l_i^2 & 1 & a^2 \\ \vdots & \vdots & \vdots \\ l_i^{N_p} & 1 & a^{N_p} \end{bmatrix} \begin{bmatrix} x_1 \\ x_2 \\ x_3 \end{bmatrix} + \begin{bmatrix} e^1 \\ e^2 \\ \vdots \\ e^{N_p} \end{bmatrix} \tag{5.32}$$

where l_i^j, a^j and e^j, $j = 1, ..., N_p$ are the N_p values of the system pole, the precalculated parameters and the approximation errors respectively, $x_1 = k_{3i}$, $x_2 = -k_{1i}k_{3i}$ and $x_3 = k_{1i} - k_{2i}$.

Equation (5.32) can be written in matrix form as:

$$\mathbf{Y} = \mathbf{M}\,\mathbf{X} + \mathbf{E} \tag{5.33}$$

In order to calculate the optimum values of X, we minimize $E^T E$ to obtain

$$\mathbf{X} = (\mathbf{M}^T\mathbf{M})^{-1}\,\mathbf{M}^T\,\mathbf{Y} \tag{5.34}$$

The desired coefficients can now be evaluated as

$$k_{3i} = x_1; \quad k_{1i} = -x_2/k_{3i}; \quad k_{2i} = k_{1i} - x_3 \tag{5.35}$$

In the case of $\lambda(i) = 5$, $\delta(i) = 1$ and for a control horizon of 15 the controller coefficients are given by:

$$\begin{aligned} l_1 &= 0.4338 - 0.6041\ \hat{a} \ / \ (1.11 - \hat{a}) \\ l_2 &= -0.4063 + 0.4386\ \hat{a} \ / \ (1.082 - \hat{a}) \\ l_3 &= -l_1 - l_2 \end{aligned} \tag{5.36}$$

These expressions give a very good approximation to the true controller parameters and fit the set of computed data with a maximum error of less than 0.6 per cent of the nominal values for the range of interest of the open loop pole.

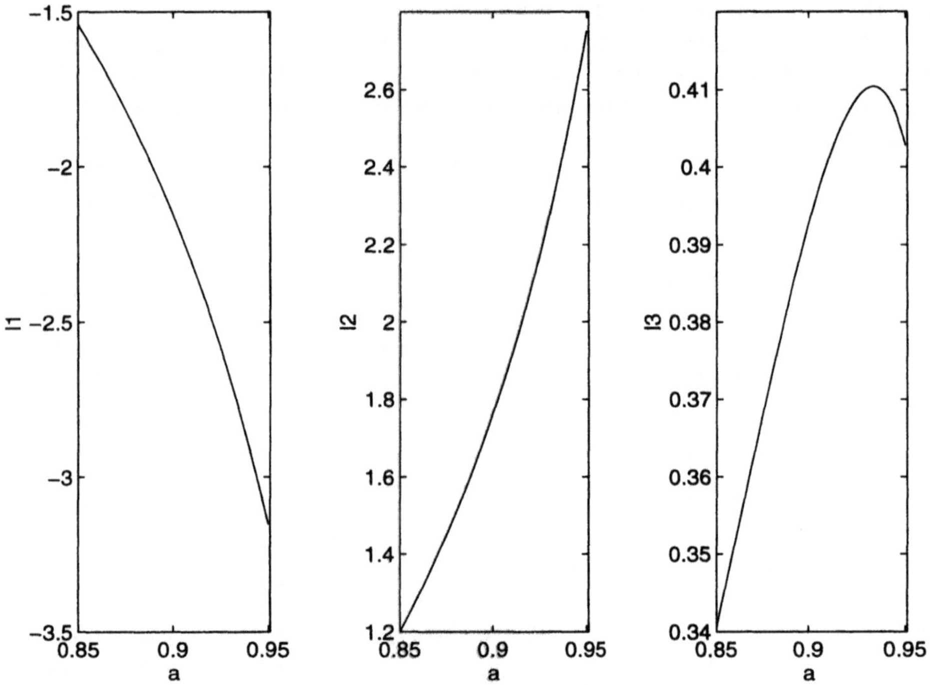

Fig. 5.3. GPC controller parameters for $\lambda = 5$

In each sampling period the adaptive controller consists of the following steps:

1. An estimation of the parameters of a linear model by measuring the inlet and outlet values of the process.
2. The adjustment of the parameters of the controller using expressions obtained for l_i (5.36).
3. Compute $\hat{y}(t + d \mid t)$ using the predictor (5.26).

4. The calculation of the control signal using (5.30).
5. The supervision of the correct working of the control.

5.3.3 Simulation studies

In this paragraph, some simulation results are shown (Figure 5.4) using data obtained from the reference test explained in chapter 3, in order to evaluate the behaviour of the controller.

As can be seen in Figure 5.4, the outlet oil temperature follows the set point and the perturbations are rejected quite rapidly. The antiresonance modes are not excited because the weighting factor has been adjusted to a value ($\lambda = 7$[1]) in order to avoid this effect (in this case, the bandwidth of the controlled system is such that antiresonance modes hardly influence the dynamic response of the controlled system, with the cost of obtaining a slower system response). If smaller weighting factors are used, oil flow changes contain medium and high frequency components near the antiresonance modes, producing oscillations in the outlet oil temperature evolution.

Notice that, in the case of the design of PID controllers in chapter 3, the effect of the antiresonance modes could be avoided by changing the desired closed loop pole placement with respect to that of the open loop system.

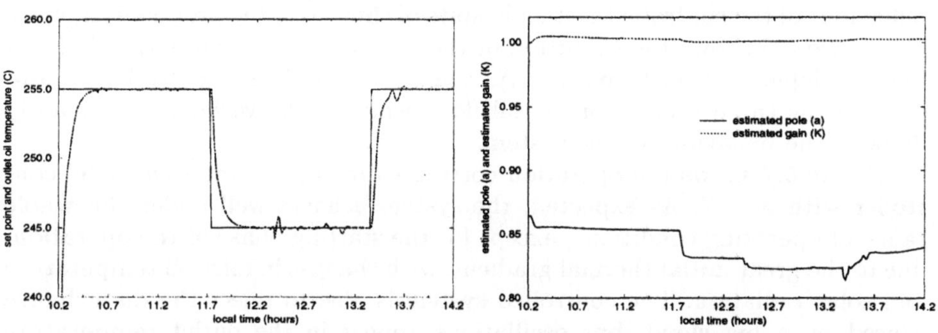

set point and outlet oil temperatures (°C) — evolution of the estimated pole and gain

Fig. 5.4. Simulation with the adaptive GPC controller - Type A model

When operating with the real plant, the objective is to obtain a response as faster as possible, trying to avoid oscillations due to the excitation of the antiresonance modes or to a wrong controller parameters tuning. As will be analyzed, due to the fact that simplified linear models have been used in the design stage of the control strategy, when a fast response is required, oscillations should appear in the outlet temperature due to unmodelled dynamics.

The disturbance rejection capabilities of this controller are similar to those of controllers analyzed in chapter 4 (notice that in all these control schema the

[1] The calculation of the adaptation law is the same expounded for $\lambda = 5$

feedforward controller in series with the plant has been used). The evolution of the estimated parameters is also similar to those controllers studied in chapter 4, under the same conditions of solar radiation and inlet oil temperature.

5.3.4 Plant results

In this section, several results obtained in the operation with the plant with the adaptive GPC are analyzed. The parameters identification mechanism has been explained in chapter 4, that is, recursive least squares incorporating data filtering and heuristic rules to supervise the behaviour of the adaptive controller.

Both the analysis of a fast adaptive GPC controller and a conservative one has been performed. In the case of the fast controller, Figure 5.5 shows the outlet oil temperature and reference temperature when controlling the plant with an adaptive GPC with a value of $\lambda = 5$. As can be seen a quite fast response is obtained (rise time of about 6 minutes). The weighting factor is increased to obtain a smoother response with less overshoot, as done in the simulation showed in previous section with a weighting factor $\lambda = 7$.

The evolution of the direct solar radiation in this test can be seen in the same figure. It corresponds to a day with scattered clouds. The oil flow changed from 4.5 l/s to 7 l/s, and the controller could maintain a good behaviour of the controlled system in spite of changes in the process dynamics.

Figure 5.6 shows the evolution of the controller parameters l_1 y l_2 (l_3 is linearly dependent on l_1 and l_2). As can be observed, the controller parameters follow the evolution of the oil flow, which is the variable that mainly dictates the behaviour of the system.

Figure 5.7 shows the operation corresponding to the adaptive GPC controller with $\lambda = 7$. As expected, the system behaves well under the whole range of operating conditions, except for the starting phase of the operation, due to the great initial thermal gradient, with changes in inlet oil temperature and solar radiation. The controlled system is able to reject the disturbance caused by a big cloud, but oscillations appear in the outlet temperature evolution due to the excitation of the antiresonance modes. This effect is manifested in Figure 5.7, in which the evolution of the estimated parameters, solar radiation, inlet oil temperature and oil flow are shown. As can be seen, the convergence of the estimated parameters within their admissible ranges to operate in an adaptive strategy is slow, mainly due to the lack of dynamic richness. In other tests, problems related to the evolution of the estimated parameters far from the prespecified range have also appear, being necessary to implement the supervisory mechanism explained in chapter 4 and to develop a robust adaptive control scheme, as done in next section.

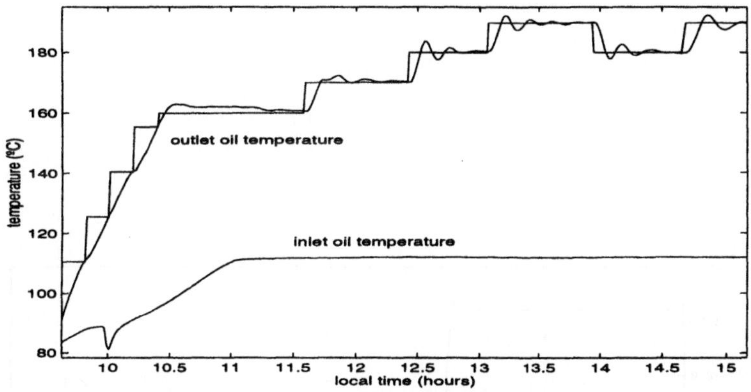

set point and outlet oil temperature (^oC)

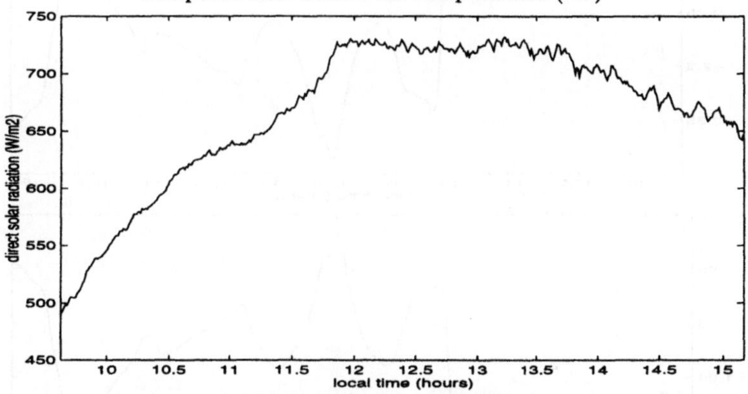

solar radiation (W/m^2)

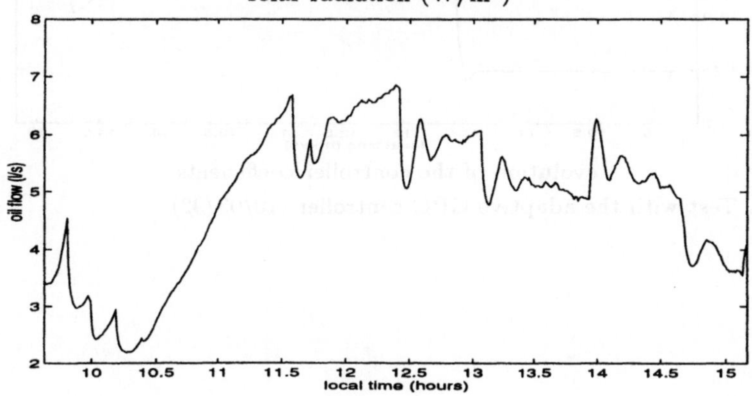

oil flow (l/s)

Fig. 5.5. Test with the adaptive GPC controller (10/03/92)

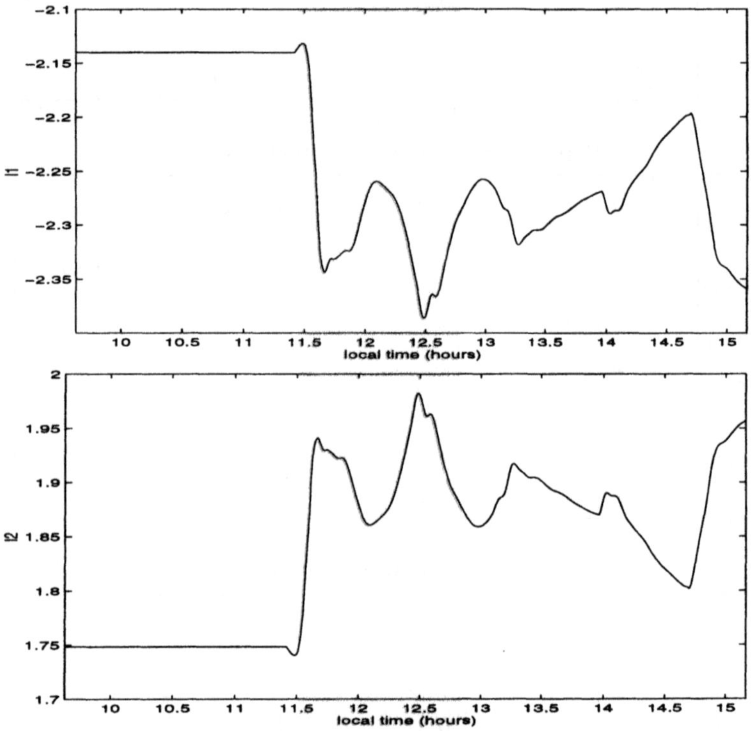

evolution of the controller coefficients

Fig. 5.6. Test with the adaptive GPC controller (10/03/92)

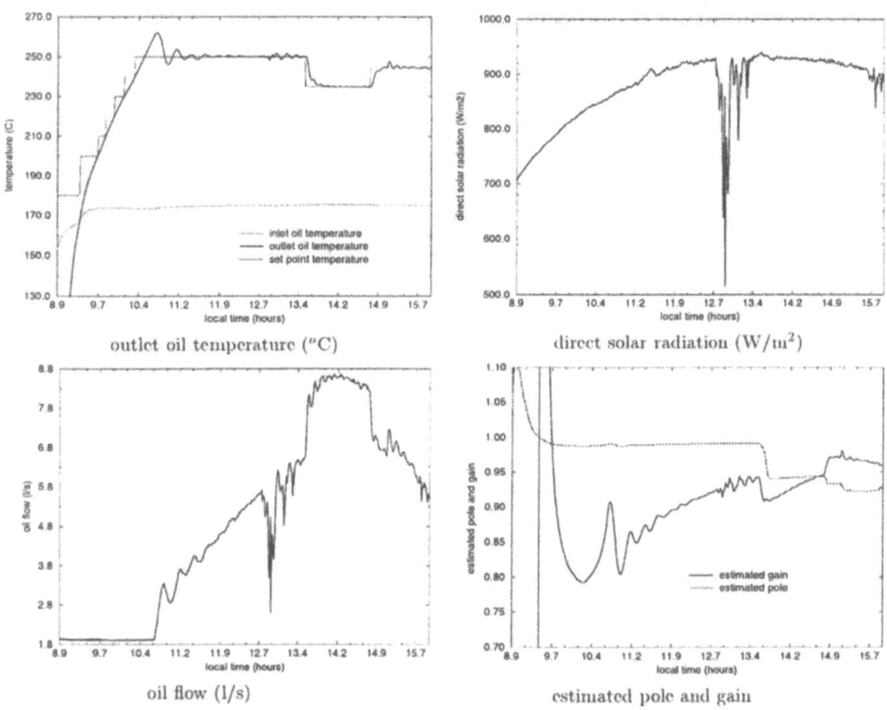

Fig. 5.7. Test with the adaptive GPC controller (02/06/95)

5.4 Robust adaptive model predictive control with bounded uncertainties

5.4.1 Introduction

This section presents the application of a robust adaptive control scheme which uses a robust identification mechanism combined with a finite horizon receding controller to cope with the process dynamics having bounded uncertainties [20].

It is a common practice to make simplifying assumptions to obtain mathematical models of real processes, especially when the models are going to be used for control purposes, where models with simple structures (linear in most cases) and sufficiently small size have to be used due to available control techniques and real time considerations. Thus, mathematical models, especially control models, can only describe the dynamics of the process in an approximative way.

Most control design techniques need a control model of the plant with fixed structure and parameters (*nominal model*), which is used throughout the design. If the control model were an exact, rather than an approximate, description of the plant and there were no external disturbances, processes could be controlled by an open loop controller. Feedback is necessary in process control because of external perturbations and model inaccuracies present in all processes. The objective of robust control is to design controllers which preserve stability and performance in spite of the model inaccuracies or uncertainties.

Although the use of feedback contemplates the inaccuracies of the model implicitly, the term *robust control* is used in literature for control systems that explicitly consider the discrepancies between the model and the real process.

There are different approaches for modelling uncertainties mainly depending on the type of technique used for designing the controllers. The most widespread types are frequency response uncertainties and transfer function parametric uncertainties. Most of the approaches assume that there is a family of models and that the plant can be exactly described by one of the models belonging to the family. That is, if the family of models is composed of linear models, the plant is also linear. The approach considered here is the one relevant to the key feature of model predictive control MPC which is to predict future values of the output variables. Uncertainties about the prediction capability of the model can be defined. The uncertainties will be considered to affect the transfer function parameters and the 1-step ahead prediction equation.

The robust controller is used in connection with a robust identification technique, which consists of determining membership sets for the parameter of the plant. Several robust identification methods have been proposed in the literature. These methods can be classified into three kinds, depending on the form of the membership sets: polyhedric, ellipsoidal and hypercubic methods.

An estimation method based on an hypercubic parameter uncertainty set will be used here [82].

The key idea of the controller here explained is that the identification algorithm determines (and progressively reduces) the uncertainty level about plant parameters. These uncertainty levels are used by a robust model predictive controller which optimizes the objective function for the worst possible case of the uncertainties.

5.4.2 Robust identification mechanism

Consider a single input – single output (SISO) process whose behaviour is dictated by the following equation:

$$y(t) = \sum_{i=1}^{n_a} a_i y(t-i) + \sum_{i=0}^{n_b} b_i u(t-d-i) + e(t) \tag{5.37}$$

where $y(t)$ and $u(t)$ are the output and input of the system respectively, at instant t; $e(t)$ is the modelling error, which is unknown but bounded (UBB); a_i and b_i are the parameters of the plant and d is the delay time.

Let us define $\boldsymbol{\varphi}(t)$ and $\boldsymbol{\theta}$ as:

$$\boldsymbol{\varphi}(t) = \begin{bmatrix} y(t-1) \\ \vdots \\ y(t-n_a) \\ u(t-d) \\ u(t-d-1) \\ \vdots \\ u(t-d-n_b) \end{bmatrix} ; \mathbf{a} = \begin{bmatrix} a_1 \\ a_2 \\ \vdots \\ a_{n_a} \end{bmatrix} ; \mathbf{b} = \begin{bmatrix} b_0 \\ b_1 \\ \vdots \\ b_{n_b} \end{bmatrix} ; \boldsymbol{\theta}^T = [\, \mathbf{a}^T \quad \mathbf{b}^T \,]$$

with $n = n_a + n_b + 1$; $\boldsymbol{\varphi}(t), \boldsymbol{\theta} \in \mathbb{R}^n$; $\mathbf{a} \in \mathbb{R}^{n_a}$; $\mathbf{b} \in \mathbb{R}^{n_b+1}$ so that equation (5.37) can be rewritten as:

$$y(t) = \boldsymbol{\varphi}^T(t)\boldsymbol{\theta} + e(t) \tag{5.38}$$

The modelling UBB errors are defined by:

$$e_{min} \leq e(t) \leq e_{max} \tag{5.39}$$

and taking into account equation (5.37) the fundamental inequality of the systems with bounded uncertainties can be written:

$$y(t) - e_{max} \leq \boldsymbol{\varphi}^T(t)\boldsymbol{\theta} \leq y(t) - e_{min} \tag{5.40}$$

This relation implies that at each sampling time t and after a new output is known, two hyperplanes are generated in $\mathbb{R}^n$, and the parameter vector $\boldsymbol{\theta}$ belongs to the region included between both hyperplanes. In this way, after k

the vertices of the hypercube $\boldsymbol{\xi}$. Other types of objective functions have been used in literature. In the work of Campo and Morari [28] a $\infty - \infty$ norm is used while in the paper of Allwright [2] a $1 - \infty$ norm is proposed.

On the other hand, if we define the following matrices:

$$\mathbf{G_a} = \begin{bmatrix} g_1 & 0 & \cdots & 0 \\ g_2 & g_1 & \cdots & 0 \\ \vdots & \vdots & \ddots & \vdots \\ g_N & g_{N-1} & \cdots & g_1 \end{bmatrix}, \ \mathbf{G_a} \in \mathbb{R}^{(N,N)} \quad , \quad g_i = \sum_{j=1}^{\min(i-1,n_a)} a_j g_{i-j} \ ,$$

$$g_1 = 1; \quad \Psi = [\ \Psi_\mathbf{y} \ \Psi_\mathbf{u} \ \mathbf{I_N}\ ] \quad ; \quad \Psi \in \mathbb{R}^{(N,n+N)}$$

The matrix $\Psi_\mathbf{y}$ depends on past output values $\{y_{k+d-1}, \ldots, y_{k+d-n_a}\}$, $\Psi_\mathbf{u}$ is a matrix depending on past and future input values $\{u_{k+N-1}, \ldots, u_k, \ldots, u_{k-n_b}\}$ and $\mathbf{I_N}$ is the identity matrix of order N. In this way, if the prediction equation (5.37) is used, then:

$$\mathbf{y} = \begin{bmatrix} y_{k+d} \\ y_{k+d+1} \\ \vdots \\ y_{k+d+N-1} \end{bmatrix} = \mathbf{G_a}\,\Psi\,\boldsymbol{\theta}^* \tag{5.46}$$

The equation (5.45) can now be written as:

$$J(\boldsymbol{\theta}^*, \mathbf{u}) = (\mathbf{G_a}\Psi\boldsymbol{\theta}^* - \mathbf{w})^T(\mathbf{G_a}\Psi\boldsymbol{\theta}^* - \mathbf{w}) + J_u \quad \text{with} \quad J_u = \lambda \sum_{j=1}^{N} (\Delta u_{k+j-1})^2 \tag{5.47}$$

The function $J(\boldsymbol{\theta}^*, \mathbf{u})$ can be expressed as a quadratic function of the parameters b_i , $i = 0, \ldots, n_b$ and the errors e_{k+d+j} , $j = 0, \ldots, N-1$, for each value of the vector $\mathbf{u}$:

$$J(\boldsymbol{\theta}^*, \mathbf{u}) = \frac{1}{2}\boldsymbol{\theta}^{*T}\mathbf{H}_{\boldsymbol{\theta}}(\mathrm{a})\boldsymbol{\theta}^* + \mathbf{q}_{\boldsymbol{\theta}}^T(\mathrm{a})\boldsymbol{\theta}^* + p_{\boldsymbol{\theta}} \tag{5.48}$$

where the matrix $\mathbf{H}_{\boldsymbol{\theta}}(\mathbf{a})$ and the vector $\mathbf{q}_{\boldsymbol{\theta}}(\mathbf{a})$ depend on the parameters a_i, $i = 1, \ldots, n_a$.

The Hessian matrix of function $J(\boldsymbol{\theta}^*, \mathbf{u})$ with respect to parameters $\mathbf{b}$ and $\mathbf{e}$ is $\mathbf{H}_{\boldsymbol{\theta}}'(\mathbf{a}) = 2(\mathbf{G_a}\Psi')^T(\mathbf{G_a}\Psi')$, which is a positive semidefinite matrix and where $\Psi' = [\Psi_\mathbf{u}\ \mathbf{I_N}]$. Since matrix $\mathbf{H}_{\boldsymbol{\theta}}'(\mathbf{a})$ is positive semidefinite, function $J(\boldsymbol{\theta}^*, \mathbf{u})$ is convex in the hypercube of the parameters $\mathbf{b}$ and $\mathbf{e}$ (Bazaraa & Shetty, theorem 3.3.8) [9], and we can assume the maximum of $J(\boldsymbol{\theta}^*, \mathbf{u})$ will be reached on one of the vertices of the hypercube of the parameters $\mathbf{b}$ and $\mathbf{e}$ (Bazaraa & Shetty, theorem 3.4.6).

For a given $\mathbf{u}$ the maximization problem is solved by determining which of the $2^{(N+n_b+1)}$ vertices of the hypercube of $\mathbf{b}$ and $\mathbf{e}$ produces the maximum value of $J(\boldsymbol{\theta}^*, \mathbf{u})$. This operation is carried out by evaluating $J(\boldsymbol{\theta}^*, \mathbf{u})$ at all the $2^{(N+n_b+1)}$ vertices of the hypercube of **b** and **e** and by looking for the maximum of $J(\boldsymbol{\theta}^*, \mathbf{u})$ at an interior point of the hypercube of **a**, due to the convexity of $J(\boldsymbol{\theta}^*, \mathbf{u})$ with respect to parameters **a** is not guaranteed. Also the maximum of $J(\boldsymbol{\theta}^*, \mathbf{u})$, at an interior point of the hypercube of **a**, can be found by applying a grid of precision p^{n_a} because of the application of a robust identification algorithm, which yields a very small hypercube at a. If a finite truncated impulse model is used the convexity of $J(\boldsymbol{\theta}^*, \mathbf{u})$ is guaranteed with respect to all the parameters [28], [2].

A method is presented in [121], which looks for the maximum of J at the vertices of the hypercube of the predictor parameters. But there are $N(N + n + 1)$ predictor parameters and therefore the hypercube of the predictor parameters has $2^{N(N+n+1)}$ vertices. On one hand, this algorithm can be too conservative because there are vertices of the hypercube of the predictor parameters that do not correspond to the vertices of the hypercube of the real parameters; and on the other hand, when N is big the amount of computation required becomes too large. When n_a is small and N big, the algorithm explained here can be faster than the above method.

The prediction vector can be written as:

$$\mathbf{y} = \mathbf{G_u}\mathbf{u} + \mathbf{f} \tag{5.49}$$

where matrix $\mathbf{G_u}$ depends on parameter vector $\boldsymbol{\theta}$, and vector **f** depends on $\boldsymbol{\theta}$ and past values. In this way, equation (5.45) can now be expressed as:

$$J(\boldsymbol{\theta}^*, \mathbf{u}) = (\mathbf{G_u}\mathbf{u} + \mathbf{f} - \mathbf{w})^T(\mathbf{G_u}\mathbf{u} + \mathbf{f} - \mathbf{w}) + J_u \tag{5.50}$$

where $J_u = \lambda(\mathbf{Mu} - \mathbf{m})^T(\mathbf{Mu} - \mathbf{m})$ with:

$$\mathbf{M} = \begin{bmatrix} 1 & 0 & 0 & \cdots & 0 \\ -1 & 1 & 0 & \cdots & 0 \\ 0 & -1 & 1 & \cdots & 0 \\ \vdots & & \ddots & \ddots & \vdots \\ 0 & 0 & \cdots & -1 & 1 \end{bmatrix}, \quad \mathbf{m} = \begin{bmatrix} u_{k-1} \\ 0 \\ \vdots \\ 0 \end{bmatrix}, \quad \mathbf{M} \in \mathbb{R}^{(N,N)}, \ \mathbf{m} \in \mathbb{R}^N$$

Thus, equation (5.50) can now be rewritten as:

$$J(\boldsymbol{\theta}^*, \mathbf{u}) = \frac{1}{2}\mathbf{u}^T\mathbf{H_u}\mathbf{u} + \mathbf{q_u}^T\mathbf{u} + p_\mathbf{u} \tag{5.51}$$

with:

$$\begin{aligned} \mathbf{H_u} &= 2(\mathbf{G_u}^T\mathbf{G_u} + \lambda\mathbf{M}^T\mathbf{M}) \\ \mathbf{q_u} &= 2[\mathbf{G_u}^T(\mathbf{f} - \mathbf{w}) - \lambda\mathbf{M}^T\mathbf{m}] \\ p_\mathbf{u} &= (\mathbf{f} - \mathbf{w})^T(\mathbf{f} - \mathbf{w}) + \lambda\mathbf{m}^T\mathbf{m} \end{aligned}$$

where $\mathbf{H_u} = \mathbf{H_u}(\boldsymbol{\theta}^*)$, $\mathbf{q_u} = \mathbf{q_u}(\boldsymbol{\theta}^*)$ and $p_\mathbf{u} = p_\mathbf{u}(\boldsymbol{\theta}^*)$. Note that $J(\boldsymbol{\theta}^*, \mathbf{u})$ is a quadratic function of **u**.

It can easily be seen that function $J^*(\mathbf{u})$ is a piece-wise quadratic function of $\mathbf{u}$. Let's divide the $\mathbf{u}$ domain $\mathbf{U}$ into different regions $\mathbf{U_p}$ so that $\mathbf{u} \in \mathbf{U_p}$ if the maximum of $J(\boldsymbol{\theta}^*, \mathbf{u})$ is attained for the vertex $\boldsymbol{\theta}_p^*$ (note that $\boldsymbol{\theta}_p^*$ is only a vertex with respect to vectors $\mathbf{b}$ and $\mathbf{e}$). For region $\mathbf{U_p}$ the function $J^*(\mathbf{u})$ is defined by:

$$J^*(\mathbf{u}) = \frac{1}{2}\mathbf{u}^T \mathbf{H_u}(\boldsymbol{\theta}_p^*)\mathbf{u} + \mathbf{q_u}(\boldsymbol{\theta}_p^*)^T \mathbf{u} + p_{\mathbf{u}}(\boldsymbol{\theta}_p^*) \tag{5.52}$$

The Hessian matrix of function $J^*(\mathbf{u})$, $\mathbf{H_u}(\boldsymbol{\theta}_p^*)$, can be assured to be positive definite by choosing a value of $\lambda > 0$. This implies that function J is strictly convex (Bazaraa & Shetty, theorem 3.3.8) [9] and that there are no local optimal solutions other than the global optimal solution (Bazaraa & Shetty, theorem 3.4.2) [9]. These results imply that the maximum function has no local optimal minimum other than the global optimal minimum. In this way, it can be assured that the solution of the MIN-MAX problem is *unique* and it is the minimum of function $J^*(\mathbf{u})$. One of the main problems of numerical optimization, the existence of local minima, is avoided and any nonlinear optimization method can be used to solve the problem.

5.4.4 Simulation studies

The model chosen to the implementation of the algorithm has been a simplified one obtained from the characterization of the step response of the plant (Type A Model). The discrete transfer function is given by:

$$g(z^{-1}) = z^{-k}\frac{bz^{-1}}{1 - az^{-1}}$$

and at the high flow level condition $k = 1$. A zero can be added to the numerator to take into account delay times non multiple of the sampling time (Type B Model). In the application to the distributed solar collector field, the series feedforward controller has been used to diminish the effects of changes in the solar radiation level and inlet oil temperature.

The performance of the proposed robust adaptive control scheme has been tested using the nonlinear computer model of the field included in chapter 2. In this case, the range of variation of the static gain of the model considered has been taken from 0.9 to 1.2 (considering the feedforward controller as a part of the controlled plant), and that of the pole of the plant from 0.8 to 0.95. An analysis of a large number of test in the real plant indicates that the maximum level of uncertainty for the UBB errors is about $e_{min} = -0.5$ and $e_{max} = 0.5$.

The design values chosen for the robust adaptive GPC controller have been: weighting factor $\lambda = 15$, prediction and control horizons $N_1 = 1$, $N_2 = N_u = 7$ (these parameters have been chosen to obtain an adequate behaviour without an excessive computational effort).

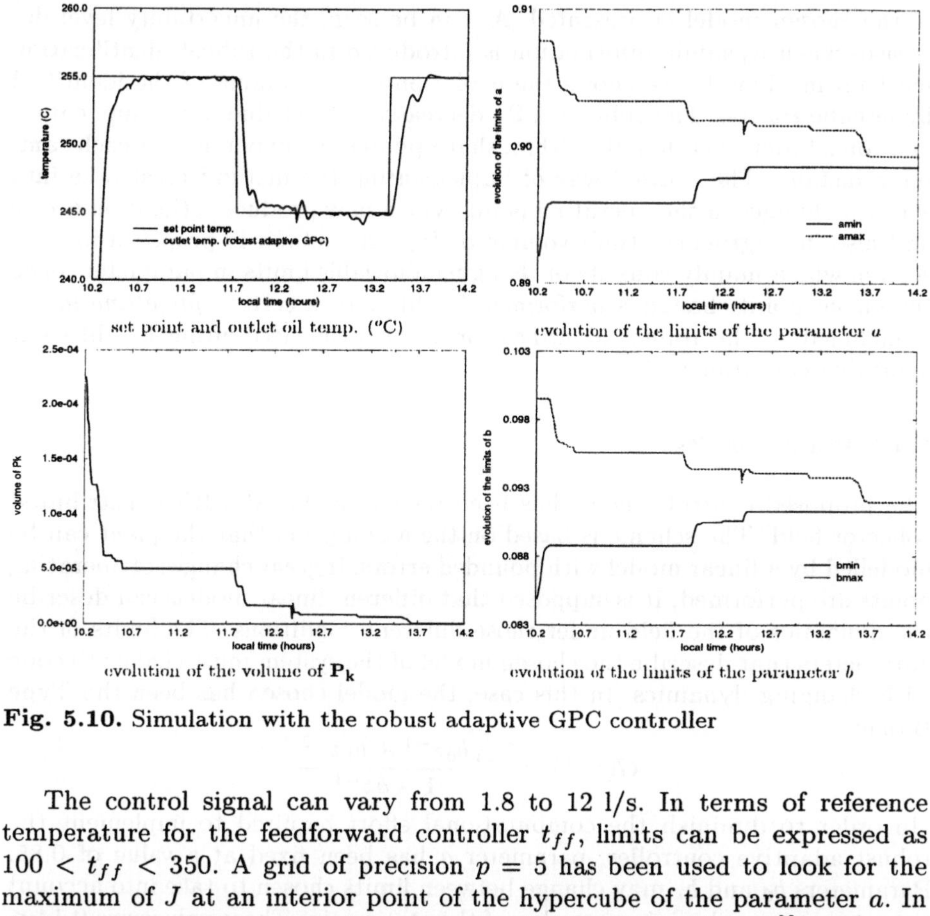

Fig. 5.10. Simulation with the robust adaptive GPC controller

The control signal can vary from 1.8 to 12 l/s. In terms of reference temperature for the feedforward controller t_{ff}, limits can be expressed as $100 < t_{ff} < 350$. A grid of precision $p = 5$ has been used to look for the maximum of J at an interior point of the hypercube of the parameter a. In this case, the number of vertices that must be tested would be $2^{N+n_b+1} \cdot p = 2^8 \cdot 5 = 1280$. Veres and Norton presented a method [121], which looks for the maximum of J at the vertices of the hypercube of the predictor parameters. But there are $N(N+n+1)$ predictor parameters and therefore the hypercube of the predictor parameters has $2^{N(N+n+1)}$ vertices. In this case, the number of vertices to be analyzed is $2^{N(N+n+1)} = 2^{70} = 1.18 \cdot 10^{21}$ is greater that those obtained with the method analyzed in this section. On one hand, this algorithm can be too conservative because there are vertices of the hypercube of the predictor parameters that do not correspond to the vertices of the hypercube of the real parameters; and on the other hand, when N is big the amount of computation required becomes too large. When n_a is small and N big, the algorithm proposed in this section can be faster than the above method.

Figure 5.10 shows a simulation of the proposed robust adaptive controller. As can be seen, the proposed robust adaptive controller behaves well in the whole range of operation. The evolution of the limits of parameters a and b

of the chosen model is presented. As can be seen, the uncertainty level decreases when dynamic information is introduced in the robust identification mechanism. This figure shows the evolution of the volume of the identified hypercube $\mathbf{P_k}$ too. The volume of $\mathbf{P_k}$ decreases in time, due to the application of a robust identification algorithm, that updates the hypercube at each sampling instant. The normal way of implementing the method must take into account changes in the operating point, which may produce estimation errors and also incongruences (null volume of $\mathbf{P_k}$). A method is proposed in next section which mainly consists of changing allowable limits in parameters each time a set point change is performed. In this way, a certain *forgetting factor* is included in the robust identification mechanism accounting for different operating conditions.

5.4.5 Plant results

The proposed control scheme has been tested on the ACUREX distributed collector field. The scheme is based on the assumption that the plant can be modelled by a linear model with bounded errors. If great changes of operating points are performed, it is supposed that different linear models can describe the dynamics of the field under those different conditions. The limits of the parameters that describe the chosen model of the system must change to cope with changing dynamics. In this case, the model chosen has been the Type B one:

$$G(z^{-1}) = z^{-1}\frac{b_0 z^{-1} + b_1 z^{-2}}{1 - az^{-1}}$$

In order to diminish the computational effort required to implement the robust adaptive controller, parameter a has been fixed at a value of 0.85. Parameters b_0 and b_1 may change between limits chosen to take into account the maximum and minimum value of the static gain. These values are $0.14 \leq b_0 \leq 0.16$, $-0.005 \leq b_1 \leq 0.005$, that produces $0.9 \leq K \leq 1.06$. Changes in the gain of the process and delay time (the zero acts as a Pade approximation of a non-integer delay time [26]) are therefore taken into account. In this case, the design values chosen are: $\lambda = 5$, $N_1 = 1$, $N = N_2 = 4$, $e_{min} = -0.5$ and $e_{max} = 0.5$. A gradient method was used to find the optimal constrained control sequence. The problem could be solved in real time because only two parameters (b_0, b_1) were identified and a small control horizon ($N = 4$) was used. Notice that to calculate $J^*(\mathbf{u})$ requires the computation of a quadratic function at $p \cdot 2^{N+n_b+1}$ points. If another b parameter were added to the list of identified parameters or if the costing horizon were increased, the number of points where the quadratic function needs to be evaluated would duplicate. This computational burden is further aggravated if the a parameters are identified. Considerable effort in reducing the computational burden and/or faster computers are required in order to make this type of controller applicable to more complex problems.

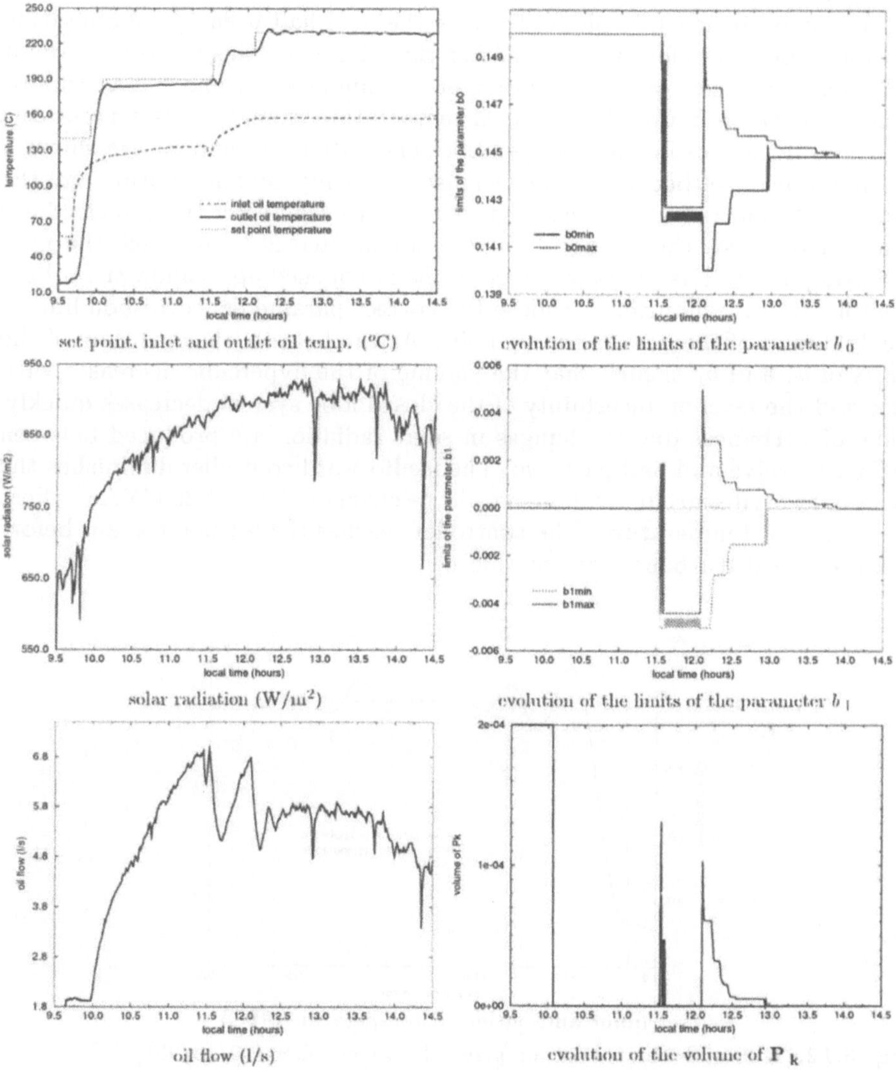

Fig. 5.11. Test with the robust adaptive GPC controller (20/10/95)

Figure 5.11 shows the evolution of the outlet temperature when controlling the plant with the proposed control scheme. The starting phase has been carried out using a fixed parameters GPC ($a = 0.85$, $b_0 = 0.15$ and $b_1 = 0$) to avoid a large overshoot. As can be seen, the response is slow and takes a long time to reach the first set point. The inlet oil temperature changes substantially during this phase, because the field had been out of operation for a long period before this test. After this phase, a change in the set point is performed, freezing the evolution of the limits of b_0 and b_1, as can be seen in the same figure. The inlet oil temperature changes at this point due to non-uniformities in the storage tank. The outlet oil temperature shows a considerable offset because of the increase of the inlet oil temperature and the increase of solar radiation according to its daily cycle. After this intermediate phase of the test, the set point value is augmented 20°C to reach the final operating point. The limits of the polytope are opened up to allow the robust identification mechanism to choose the correct parameters corresponding to the dynamics of the new operating point. As can be seen, the evolution of the limits of b_0 and b_1 is such that the volume of the hypercube decreases with time and the level of uncertainty of the closed loop system decreases quickly. Some disturbances due to changes in solar radiation are produced between half past twelve and half past two. The feedforward controller diminishes the effect of these disturbances, but some large changes of about 200 W/m^2 affect the outlet oil temperature. The controller reaches the set point again before the next big disturbances are produced.

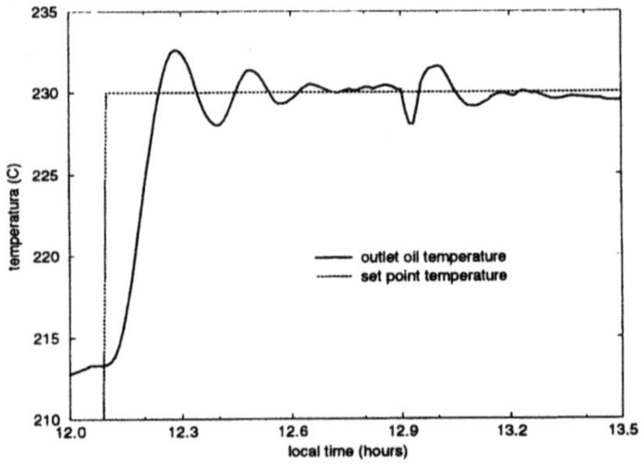

set point and outlet oil temperatures (°C)

Fig. 5.12. Test with the robust adaptive GPC controller (20/10/95)

Figure 5.12 shows the evolution at the last change of the set point. As can be seen, the robust identification algorithm updates the values and limits of the parameters to cope with the dynamics corresponding to the new operating point. The evolution of the outlet oil temperature after a change of the set point of 20°C shows very little overshoot (less than 3°C) and zero offset.

Oscillations are due to passing clouds. The rise time is about 8 minutes, which corresponds to a very acceptable response for the considered process.

Disturbance rejection

In this case, a plot showing variations of the outlet oil temperature due to changes in solar radiation of about 200 W/m^2 is presented. Oscillations due to this variations in solar radiation are less than 4°C. In this case the robust identification algorithm is less sensible to disturbances (the operation is practically performed with a fixed controller around its design point when the hypercube has closed) than any usual adaptive control scheme.

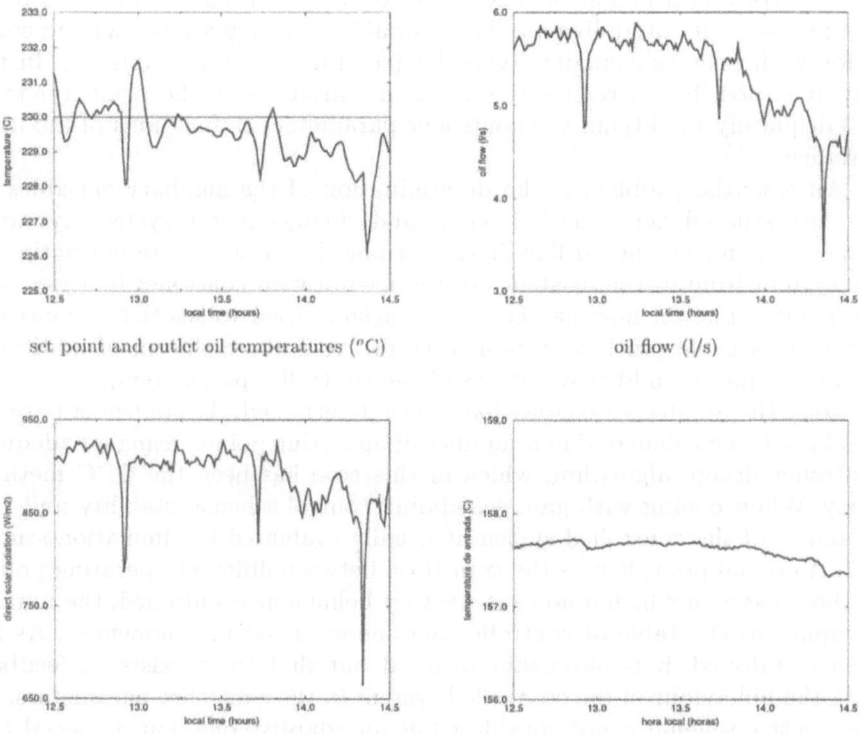

Fig. 5.13. Disturbance rejection with the robust adaptive GPC controller (20/10/95)

5.5 Gain scheduling generalized predictive control

5.5.1 Introduction

In many situations it is known how the dynamics of a process change with the operating conditions. It is then possible to change the controller parameters taking into account the actual operating point of the system. Gain scheduling is a control scheme with open loop adaptation, which can be seen as a feedback control system in which the feedback gains are adjusted by a feedforward compensation. Gain scheduling control is a nonlinear feedback of a special type: it posses a linear controller which parameters are modified depending on the operating conditions in a prespecified manner.

The working principle of this kind of controllers is simple, and it is based on the possibility of finding auxiliary variables which guarantee a good correlation with process changing dynamics (see Figure 1.3 in chapter 1). In this way, it is possible to reduce the effects of variations in the plant dynamics by adequately modifying the controller parameters as functions of auxiliary variables.

An essential problem is the determination of the auxiliary variables. In the case studied here, the behaviour and changes in the system dynamics mainly depend on the oil flow if very strong disturbances are not acting on the system (due to the existence of the feedforward controller in series with the plant). The oil flow has been the variable used to select the controller parameters table (with a certain filter to avoid the inclusion of additional dynamics due to sudden variations of the controller parameters).

Once the auxiliary variables have been determined, the controller parameters have to be calculated in a number of operating points, using an adequate controller design algorithm, which in this case has been the GPC methodology. When coping with gain scheduling control schema, stability and performance of the controlled system is usually evaluated by simulation studies [87]. A crucial point here is the transition between different operating points. In those cases in which a non-satisfactory behaviour is obtained, the number of inputs to the table of controller parameters must be augmented. As has been mentioned, it is important to point out that there exists no feedback from the behaviour of the controlled system to the controller parameters. So, this control scheme is not considered as an adaptive one, but a special case of a nonlinear controller.

The main disadvantages of gain scheduling controllers are:

- It is an open loop compensation: there is no way to compensate for a wrong election of the controller parameters within the table.
- Other important inconvenient is that the design stage of the strategy often consumes too much time and effort. The controller parameters must be calculated for a sufficient number of operating points, and the behaviour of the controlled system has to be checked in very different operating conditions.

Its main advantage consists of the easiness in changing controller parameters in spite of changes in process dynamics. As classical examples of applications of this kind of controllers, the following control fields can be mentioned: design of ship steering autopilots, pH control, combustion control, engine control, design of flight autopilots, etc. [6].

5.5.2 Plant models and fixed parameter controllers

As has been mentioned in chapter 2, the frequency response of the plant has been obtained by performing PRBS test in different operating conditions both using the plant and the nonlinear distributed parameter model. In this way, different linear models were obtained from input-output data in different working conditions. These models relate changes in oil flow to those of the outlet oil temperature, and can take into account the antiresonance characteristics of the plant if they are adequately adjusted. The control structure proposed is shown in Figure 5.14. As can be seen, the output of the generalized predictive controller is the input (`trff`) of the series compensation controller, which also uses the solar radiation, inlet oil temperature and reflectivity to compute the value of the oil flow which is sent to the pump controller.

The controller parameters were obtained from a linear model of the plant. From input-output data of the plant, the degrees of the polynomials A and B and the delay (of a CARIMA plant model) that minimizes Akaike's Information Theoretic Criterion (AIC) were found to be $n_a = 2$, $n_b = 8$ and $d = 0$. By a least squares estimation algorithm, the following polynomials were obtained using input-output data of one test with oil flow around 6 l/s:

$$A(z^{-1}) = 1 - 1.5681z^{-1} + 0.5934z^{-2}$$

$$B(z^{-1}) = 0.0612 + 0.0018z^{-1} - 0.0171z^{-2} + 0.0046z^{-3}$$

$$+ 0.0005z^{-4} + 0.0101z^{-5} - 0.0064z^{-6} - 0.015z^{-7} - 0.0156z^{-8}$$

The most adequate value for the control horizon ($N = 15$) has been calculated taking into account the values of the fundamental time constant and the sampling period used for control purposes. In this case, $N_1 = 1$ and $N_2 = 15$. The value of λ was determined by simulation studies using the nonlinear model and was found to be $\lambda = 6$ (fast) and $\lambda = 7$ (without overshoot). For smaller values of λ, faster and more oscillatory responses were obtained. Following the design procedure of the GPC methodology, the controller parameters corresponding to $\lambda = 7$ were obtained (Table 5.1).

field can be seen. The curves shown in Figure 5.16 were obtained by a spectral analysis of the input-output signals of the model at different operating points (PRBS signals were used for the input).

As can be seen, the frequency response changes significantly for different operating conditions. The steady state gain changes for different operating points, as well as the location of the antiresonance modes.

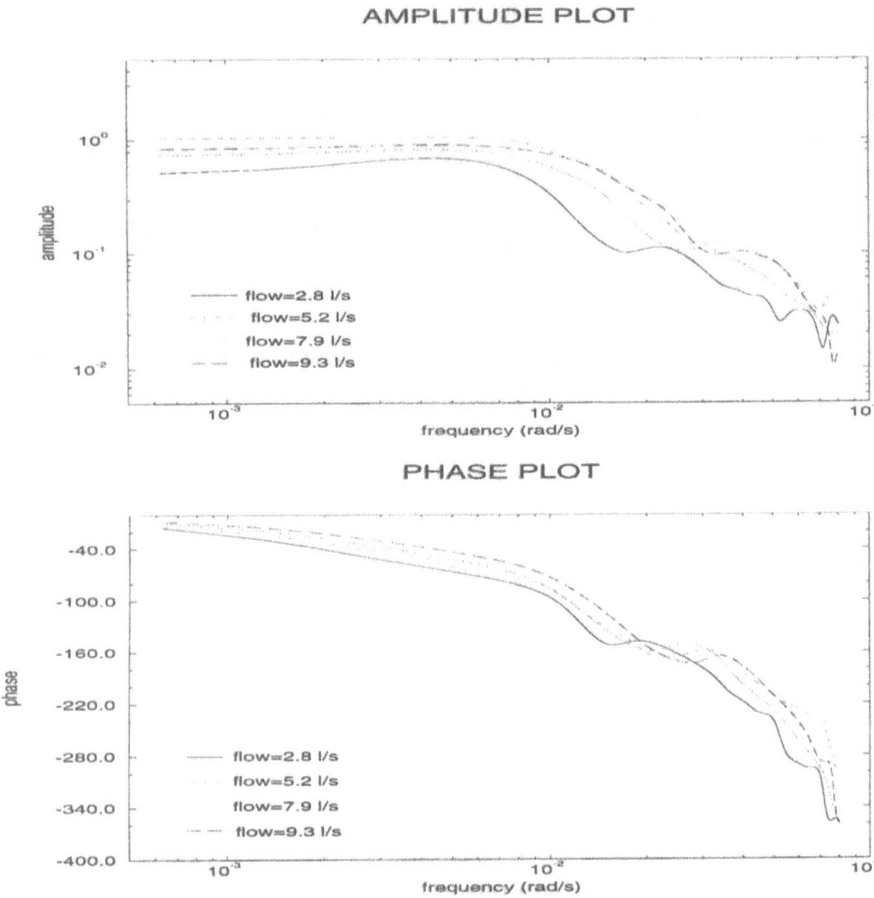

Fig. 5.16. Frequency response of the field in different operating conditions

Taking into account the frequency response of the plant and the different linear models obtained from it, it is clear that a self-tuning controller based on this type of model is very difficult to implement. The fundamental reason is the fact that the estimation of the model parameters requires a lot of computation when the number of estimated parameters increases and the convergence of the estimation process is seldom accurate and fast enough.

In order to check the influence of the parameter estimation on the controller performance, changes of ±5 percent with respect to its nominal values in each one of the estimated parameters of the high order linear models (with three parameters in the denominator and nine in the numerator) obtained from the frequency response of the plant, maintaining the other at their nominal values were produced, resulting in variations of the values of the controller coefficients of nearly ±30 percent in some cases (the controller parameters are mainly dependent on the coefficients of polynomial $A(z^{-1})$). This can be a drawback especially in cases in which dynamic information of the input signal is poor for estimation purposes, and the estimated parameters are far from their appropriate values.

Other way of coping with changing dynamics is by using a gain scheduling controller, by making the controller parameters dependent on some variables which indicate the operating conditions.

5.5.3 Gain scheduling control of the distributed solar collector field

With the input-output data used to obtain the frequency responses shown in Figure 5.16 and using the method and type of model described previously for the case of a high order fixed parameter controller, process ($a[i]$ and $b[i]$) and controller ($l[i]$) parameters were obtained for several oil flow conditions (q_1 = 2.8 l/s, q_2 = 5.2 l/s, q_3 = 7.9 l/s and q_4 = 9.3 l/s), using different values of the weighting factor λ. Tables 5.2 and 5.3 contain model and control parameters respectively for a weighting factor $\lambda = 6$. A value of $\lambda = 7$ has also been used to obtain responses without overshoot.

Table 5.2. Coefficients of polynomials $A(z^{-1})$ and $B(z^{-1})$

oil flow (l/s) / **model param.**	q_1	q_2	q_3	q_4
$a[1]$	-1.7820	-1.438	-1.414	-1.524
$a[2]$	0.81090	0.5526	0.5074	0.7270
$b[0]$	0.00140	0.0313	0.0687	0.0820
$b[1]$	0.03990	0.0660	0.0767	0.0719
$b[2]$	-0.0182	-.0272	-.0392	-.0474
$b[3]$	-0.0083	0.0071	0.0127	0.0349
$b[4]$	0.00060	0.0118	0.0060	0.0098
$b[5]$	-.00001	0.0138	-.0133	-.0031
$b[6]$	0.00130	0.0098	-.0156	0.0111
$b[7]$	0.00160	0.0027	-.0073	0.0171
$b[8]$	0.00450	-.0054	0.0037	0.0200

The controller parameters which are applied in the real operation are obtained by using a linear interpolation with the data given in Table 5.3. It

Table 5.3. GPC controller coefficients in several operating points ($\lambda = 6$)

oil flow (l/s) **controller coef.**	q_1	q_2	q_3	q_4
l[0]	-7.0481	-1.4224	-1.1840	-1.3603
l[1]	16.2223	3.84390	3.48440	3.02280
l[2]	-9.5455	-2.7794	-2.6527	-2.0142
l[3]	0.03910	-0.0139	0.00860	0.03740
l[4]	0.00980	0.00830	-0.0184	0.02730
l[5]	0.00560	0.02610	-0.0352	0.01080
l[6]	-0.0070	0.03390	-0.0239	-0.0197
l[7]	-0.0016	0.02480	0.02630	0.00460
l[8]	-0.0793	0.00880	0.03980	0.05070
l[9]	-0.1575	-0.0822	-0.0869	-0.1098
l[10]	0.36470	0.16410	0.19600	0.12480
l[11]	0.82620	0.83010	0.89360	0.87390
l[12]	0.37130	0.35800	0.35230	0.35170

is important to point out that to avoid the injection of disturbances during the controller gains adjustment, it is necessary to use a smoothing mechanism of the transition surfaces of the controller gains. In this case, a linear interpolation in combination with a first order filter has been used, given a modified flow $Q(t) = .95\ Q(t-1) + .05\ q(t)$ (being $q(t)$ the value of oil flow in instant t and $Q(t)$ the filtered value used for controller parameters adjustment). The linear interpolation has also been successfully applied by [55]. Other kind of gain scheduling approach can be obtained by switching from one controller parameters to others depending on the flow conditions, without interpolating between controller parameters. The set of controller parameters c can be obtained by choosing between one of the sets c_i in table 5.3, related to flow conditions q_i $(i = 1, 2, 3, 4)$:

$$\begin{aligned} \text{if } \frac{q_{i-1}+q_i}{2} < q \le \frac{q_i+q_{i+1}}{2} \quad & \text{then} \quad c = c_i,\ i = 2,3 \\ \text{if } q \le q_1 \quad & \text{then} \quad c = c_1 \\ \text{if } q \ge q_4 \quad & \text{then} \quad c = c_4 \end{aligned}$$

The control structure is similar to the one obtained for the fixed controller previously studied. The optimal realization of the gain scheduling controller consists of calculating the controller parameters in a number of operating conditions, and suppose that the values of the controller coefficients are constant between different operating conditions, generating a control surface based on an optimization criterion which takes into account the tracking error and control effort. It is evident that if the procedure is applied at many working points, an optimum controller will be achieved for those operating conditions if there is a high correlation between the process dynamics and the auxiliary variable. The problem of this solution is that the design process become tedious. This is one of the main reasons to include a linear interpolation between the controller parameters.

5.5.4 Plant results

In the case of real tests, similar results were obtained and depending on the operating point, disturbances due to passing clouds, inlet oil temperature variations, etc., different performance was achieved.

Figure 5.17 shows the results of one of these tests with the gain scheduling GPC with $\lambda = 7$. The operating conditions correspond to a clear afternoon with the solar radiation changing from 800 W/m^2 to 660 W/m^2 and oil flow changing from 3.75 l/s to 2 l/s. As can be seen, the effect of the antiresonance modes does not appear in the response, due to the use of an extended high order model with accounts for these system characteristics.

Figure 5.18 shows the result of a test with a weighting factor $\lambda = 7$ corresponding to a day of intermittent scattered clouds which produce large changes in the solar radiation level and inlet oil temperature changing from 170°C to 207°C. As can be seen, the outlet oil temperature follows the set point in spite of changing operating conditions and the high level of noise in the radiation level produced by clouds.

The results of a test corresponding to a day with sudden changes in the solar radiation caused by clouds can be seen in Figure 5.19. As can be seen the controller (also designed with $\lambda = 7$) is able to handle different operating conditions and the sudden perturbations caused by the clouds.

After the presented test using a weighting factor $\lambda = 7$, two new test campaigns were carried out to test the behaviour of the controller with a weighting factor $\lambda = 6$. In the first campaign, the evaluation of the controller performance was considered. In the second, the behaviour of the controller operating in extreme working conditions was studied.

Figures 5.20 and 5.21 show the results obtained in the first campaign (March, 24-25, 1994). As can be seen, a fast response is obtained, with a good performance in different operating conditions. The initial oscillation showed in Figure 5.20 corresponds to the perturbation caused by bringing into service one of the loops initially closed, diminishing momentarily the oil flow which circulated through the rest of the loops, an concretely, through the control loop (that with the highest temperature in each sampling time).

As has been mentioned, during the second test campaign the system was forced to operate in extreme conditions. Figure 5.22 shows the result obtained when operating without changes in the inlet oil temperature values but with high clouds which produce sharp peaks in the solar radiation evolution. As can be seen, during a long part of the operation the set point was let to a fixed value.

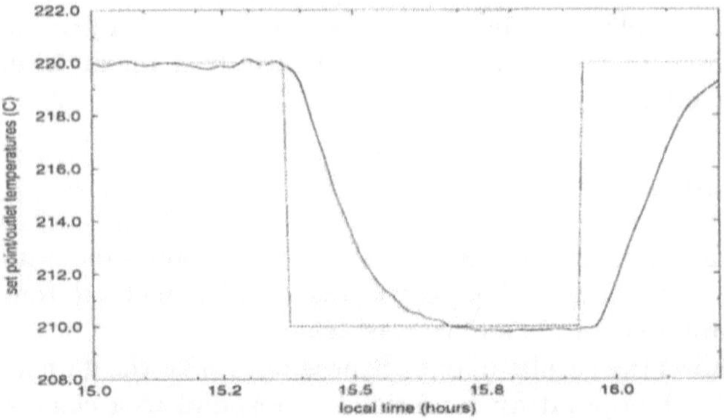

set point and oil outlet temperatures ^{o}C

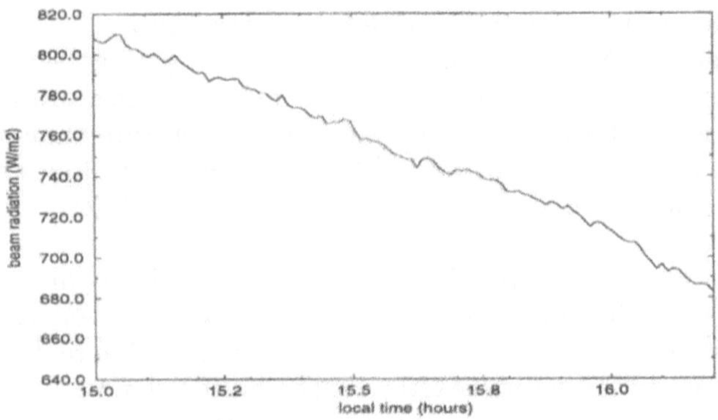

direct solar radiation (W/m^2)

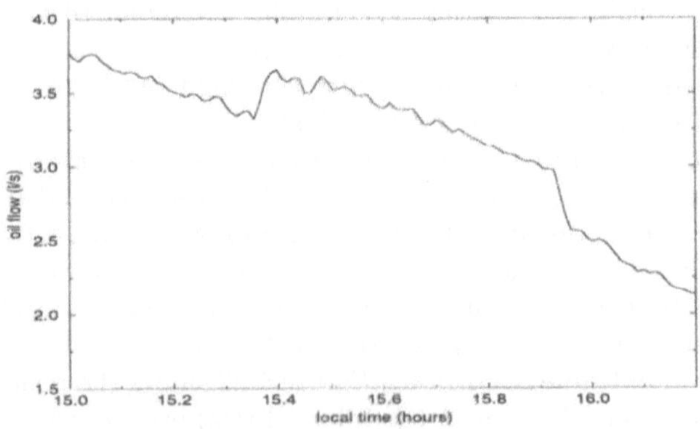

oil flow (l/s)

Fig. 5.17. Test with the gain scheduling GPC controller. $\lambda = 7$ (15/03/93)

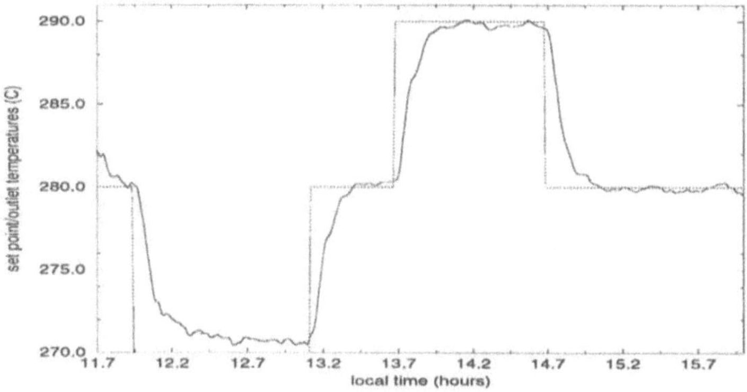

set point and outlet oil temperature (^oC)

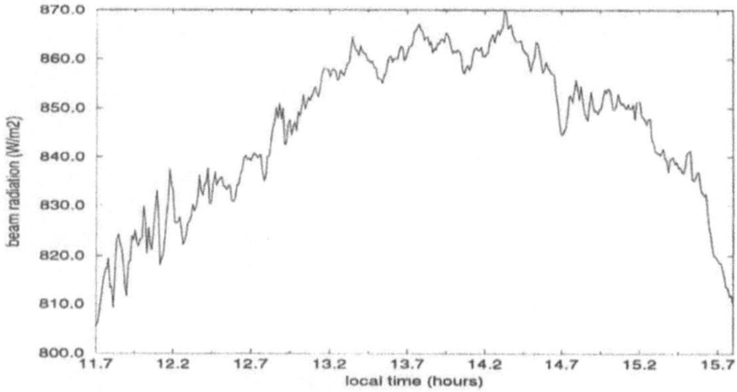

direct solar radiation (W/m^2)

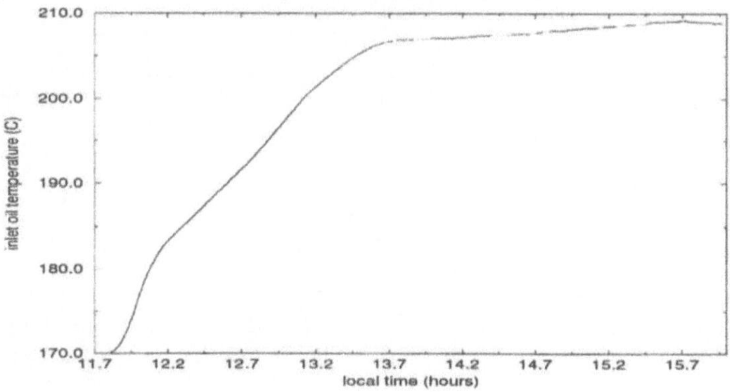

inlet oil temperature (^oC)

Fig. 5.18. Test with the gain scheduling GPC controller. $\lambda = 7$ (19/03/93)

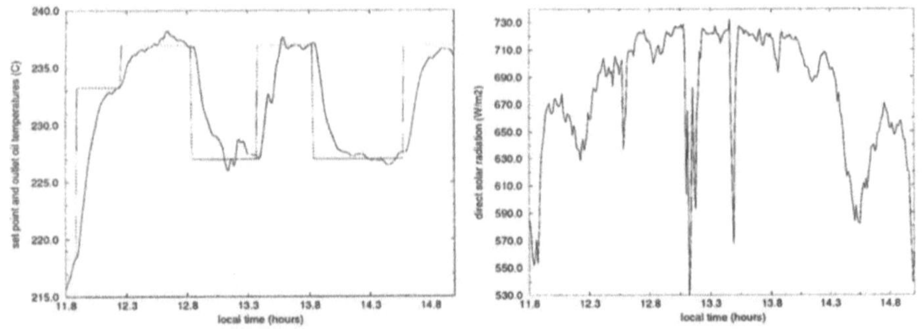

set point and outlet oil temperatures (°C) direct solar radiation (W/m^2)

Fig. 5.19. Test with the gain scheduling GPC controller. $\lambda = 7$ (21/03/93)

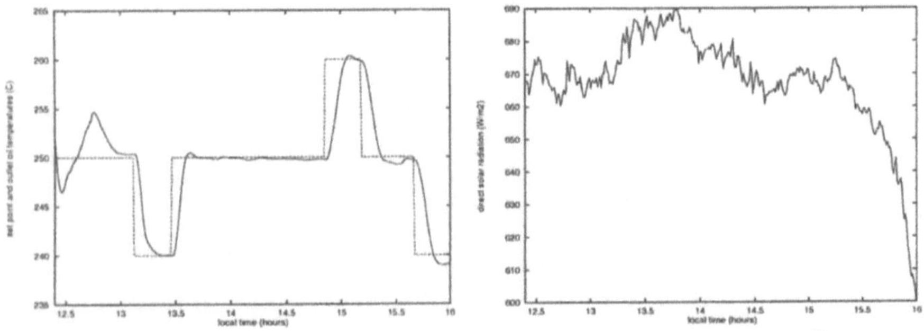

set point and outlet oil temperatures (°C) direct solar radiation (W/m^2)

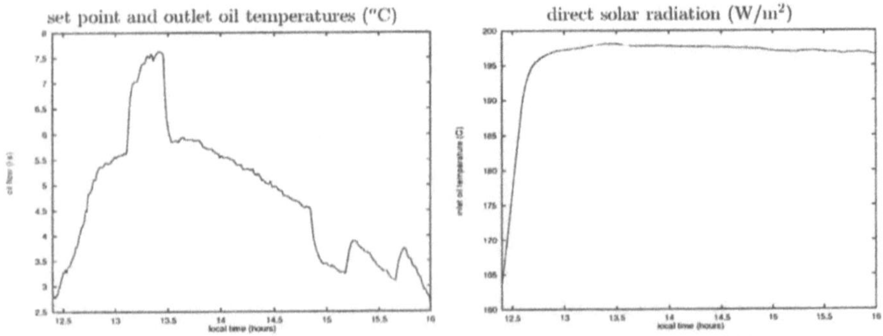

oil flow (l/s) inlet oil temperature (°C)

Fig. 5.20. Test with the gain scheduling GPC controller. $\lambda = 6$ (24/03/94)

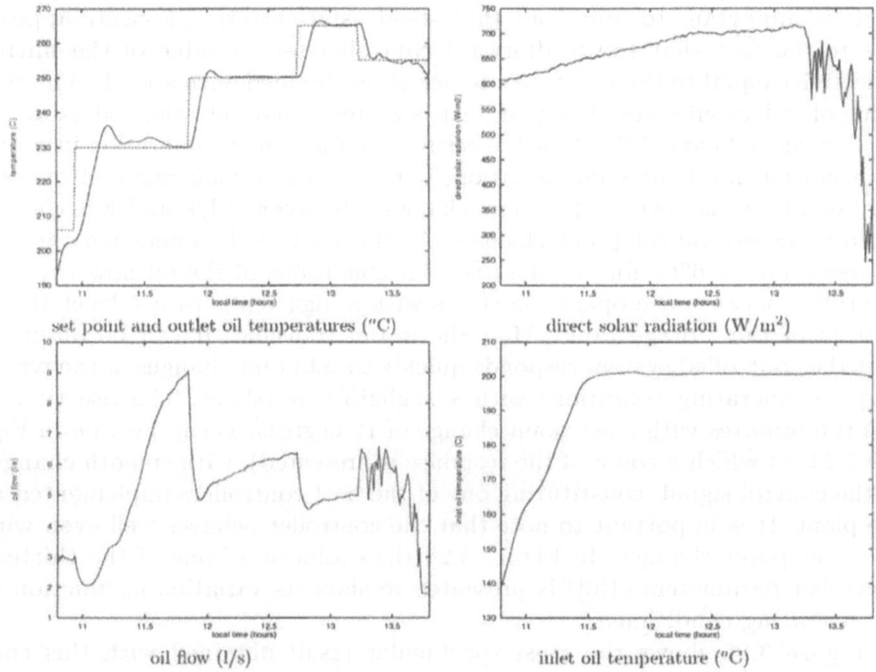

set point and outlet oil temperatures (°C) direct solar radiation (W/m²)

oil flow (l/s) inlet oil temperature (°C)

Fig. 5.21. Test with the gain scheduling GPC controller. $\lambda = 6$ (25/03/94)

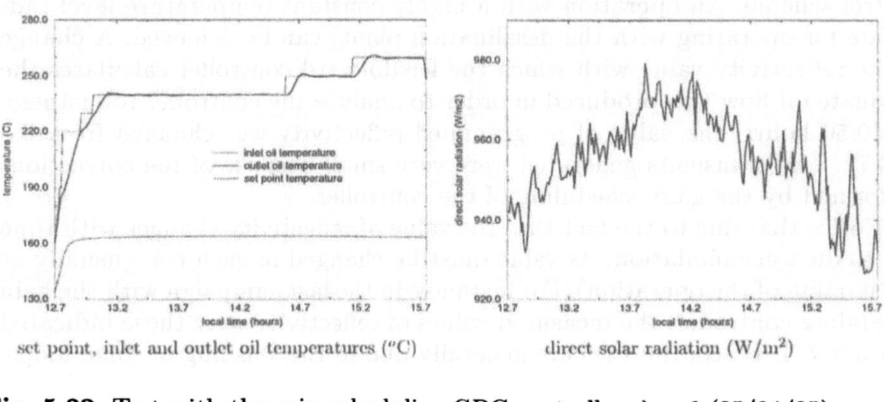

set point, inlet and outlet oil temperatures (°C) direct solar radiation (W/m²)

Fig. 5.22. Test with the gain scheduling GPC controller. $\lambda = 6$ (25/04/95)

In this part, the inexistence of steady state errors (as expected using a CARIMA model) and good disturbance rejection properties could be checked.

It is important to point out that steady state errors can easily appear, due to the fact that the feedforward controller uses a value of the mirror reflectivity equal to the last measure, which is obtained once a week. The real value of reflectivity can change if dust is accumulated onto the collectors.

Figures 5.23 and 5.24 show the results obtained in the operation in a day with normal levels of solar radiation, but in which a wide range of operating conditions is covered (oil flow changing between 2 l/s and 8.8 l/s) by performing several set point changes. At the start of the operation there is an overshoot of 6°C, due to the irregular conditions of the oil flow through the tubes because the operation starts with a high temperature level at the bottom of the storage tank. After the initial transient, it can be observed that the controlled system responds quickly to set point changes in the whole range of operating conditions with a negligible overshoot. The rise time is about 6 minutes with a set point change of 15 degrees, as can be seen in Figure 5.23 (in which a zoom of the response is presented), with smooth changes in the control signal, constituting one of the best controllers implemented at the plant. It is important to note that the controller behaves well even with great set point changes. In Figure 5.24 the evolution of one of the thirteen controller parameters (l[0]) is presented to show its variation in function of the operating conditions.

Figure 5.25 shows the most spectacular result obtained with this controller. In this test, the solar desalination plant was connected to the field. The oil coming from the desalination plant produced a decrease in the values of the temperature of the oil entering the field. In this case, a diminish of 60°C was produced, and the controller was able to cope with it. The outlet oil temperature suffered an a oscillation of less than 10°C Moreover, the solar radiation changes considerably and this disturbance was rejected by the control scheme. An operation with a highly constant temperature level (adequate for operating with the desalination plant) can be achieved. A change in the reflectivity value with which the feedforward controller calculates the adequate oil flow was produced in order to analyze the controller robustness. At 10:50 hours, the value of programmed reflectivity was changed from 0.8 to 0.72. The transients generated were very small because of the corrections performed by the gain scheduling of the controller.

Notice that due to the fact that the value of reflectivity changes with time due to dust accumulation, its value must be changed in each test (usually at the starting of the operation). For instance, in the last campaign with the gain scheduling controller, the measured values of reflectivity were those indicated in Table 5.4. Discontinuities are generally due to the washing of some loops.

Table 5.4. Values of reflectivity R

date	24/4/95	2/5/95	8/5/95	15/5/95	22/5/95	12/6/95	19/6/95
R	0.73	0.826	0.829	0.741	0.813	0.719	0.883

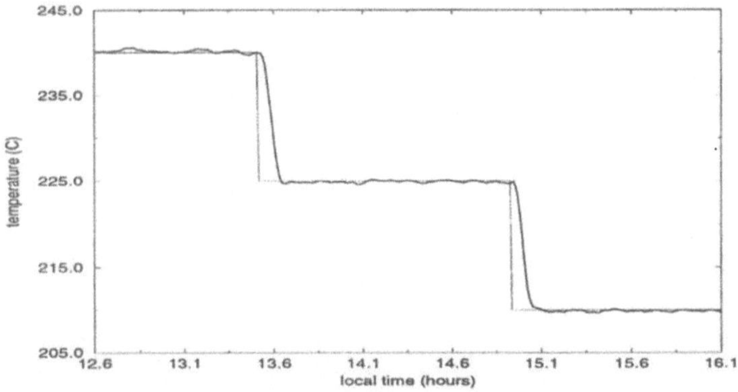

set point and outlet oil temperature (^oC)

Fig. 5.23. Test with the gain scheduling GPC controller. $\lambda = 6$ (29/05/95)

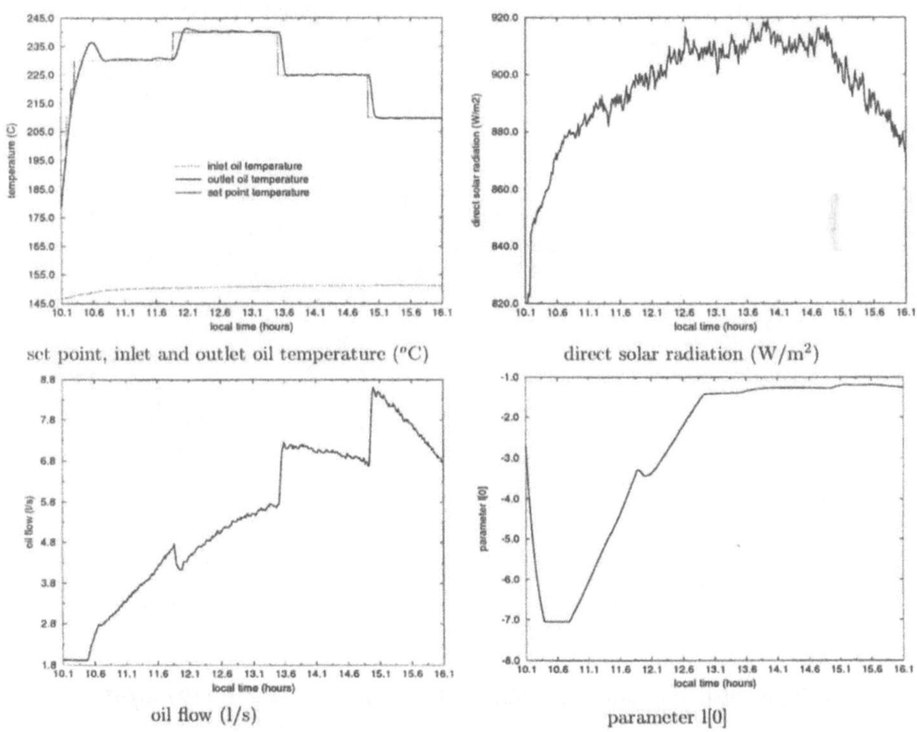

set point, inlet and outlet oil temperature (^oC)

direct solar radiation (W/m^2)

oil flow (l/s)

parameter l[0]

Fig. 5.24. Test with the gain scheduling GPC controller. $\lambda = 6$ (29/05/95)

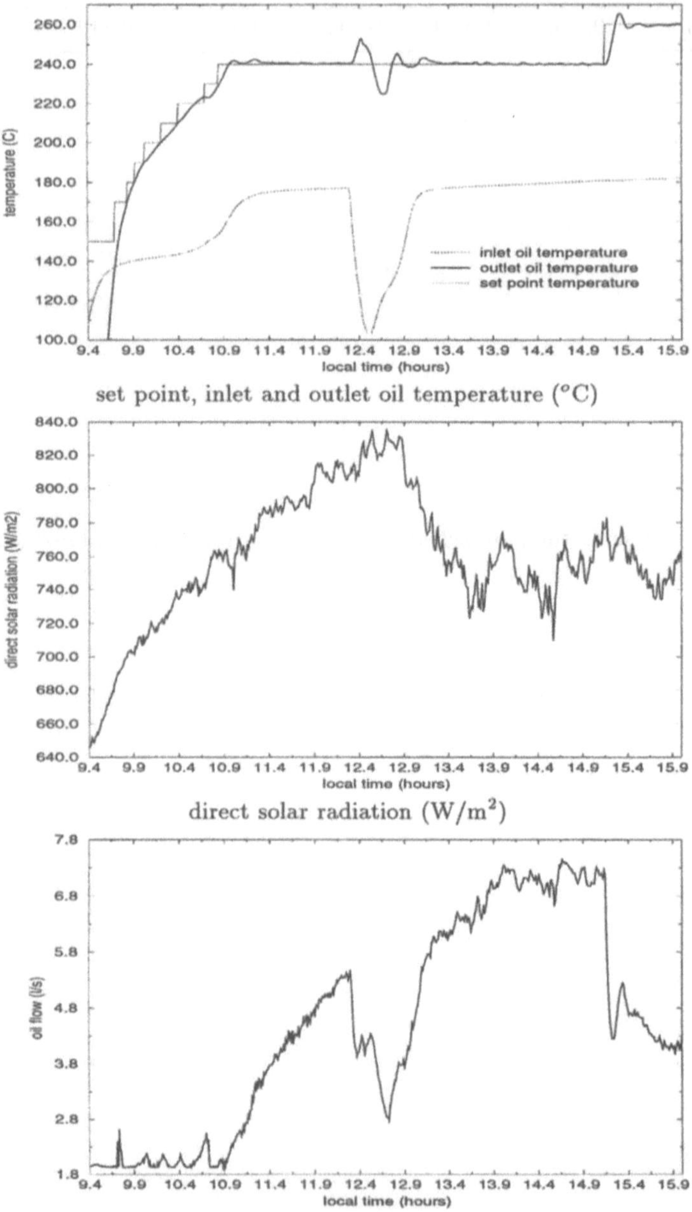

Fig. 5.25. Test with the gain scheduling GPC controller. $\lambda = 6$ (12/06/95)

Disturbance rejection

The gain scheduling GPC controller has demonstrated to have a very good behaviour not only in respect to set point tracking but also in disturbance rejection capabilities. In Figure 5.25 the behaviour under strong disturbances has been shown. In Figure 5.26 the result obtained when operating with scattered passing clouds is presented (with variations in direct solar radiation of 400 W/m^2). In spite of such disturbances, the controlled system response presents oscillations around the set point less than 3°C, which constitutes a splendid result.

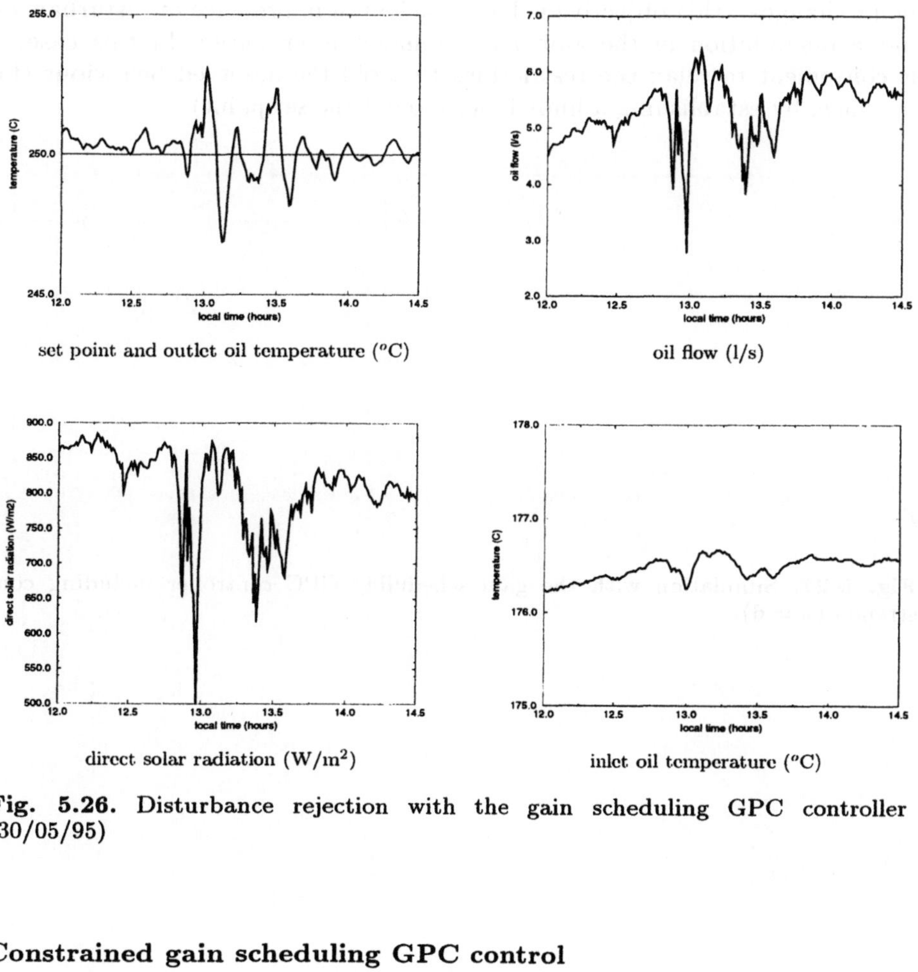

set point and outlet oil temperature (°C) — oil flow (l/s)

direct solar radiation (W/m^2) — inlet oil temperature (°C)

Fig. 5.26. Disturbance rejection with the gain scheduling GPC controller (30/05/95)

Constrained gain scheduling GPC control

To analyze the effect of constraints in the control problem, the algorithm explained at the beginning of the chapter has been implemented. It is important to point out that a control algorithm accounting for constraints is very useful to provide a secure operation, but must be used carefully to avoid the degradation of the performance when coping with systems with strong disturbances. The reference test has again be used to verify these comments.

Figure 5.27 also shows the response (outlet temperature) obtained with the gain scheduling GPC controller with $\lambda = 6$, with and without constraints in the control signal. In the unconstrained case, the effect of a cloud at the medium part of the operation is negligible. If the same controller is used but including control signal restrictions (flow saturated between 2 and 12 l/s) and outlet oil temperature two-side overshoot restrictions (the outlet temperature never exceeds the reference both in positive and negative set point changes), this objective is hardly achieved in presence of disturbances, but a degradation in the global performance is produced. In this case, it is convenient to relax the restrictions to avoid the observed behaviour (for instance, by establishing a limit band around the set point).

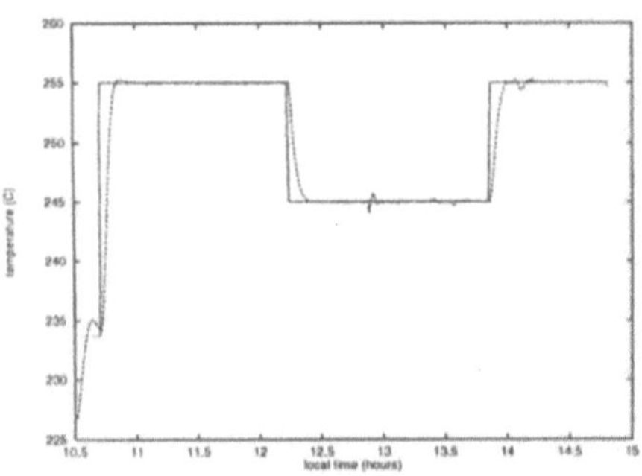

set point and outlet oil temperatures (°C) without constraints

set point and outlet oil temperatures (°C) with constraints

Fig. 5.27. Simulation with the gain scheduling GPC controller including constraints ($\lambda = 6$)

5.6 GPC scheme with nonlinear prediction of the free response

The distributed collector field is a nonlinear dynamic process. Model predictive control of nonlinear processes leads to nonlinear and usually non-convex optimization problems which are computationally very demanding. This section describes a method developed by some of the authors which makes use of the fact that although the plant characteristics change from operating point to operating point, linearity can be assumed in the neighborhood of a particular operating point. The GPC control algorithm is modified to include a kind of *free response* obtained from the nonlinear model at the optimization stage of the algorithm. This control strategy allows for including the effect of measurable disturbances within a GPC framework [19]. This scheme does not use the feedforward controller developed in chapter 3, but includes the disturbance dynamics within a nonlinear process model used for prediction purposes.

5.6.1 Nonlinear GPC scheme

The proposed control structure can be seen in Figure 5.28. A nonlinear model of the plant is used to generate the free response of the plant due to past control actions and past and future disturbances (calculated by using a convenient predictor) and considering the control signal as being at the last value. Notice that the predictive nature of the GPC and the availability of a model of the process make it possible to incorporate the dynamic disturbances caused by changes in solar radiation and inlet oil temperature into a GPC framework. A linear incremental model is used to generate the forced response. The incremental control actions are obtained from the linear model by minimizing:

$$J(N_1,N_2,N_u) = E\{\sum_{j=N_1}^{N_2} [\hat{y}_{free}(t+j \mid t) + \sum_{i=N_1}^{i=j} \Delta\hat{y}_{forced}(t+i \mid t) - w(t+j)]^2$$

$$+\sum_{j=1}^{N_u} \lambda(j)[\Delta u(t+j-1)]^2\}$$

where $\Delta u(t+j-1)$ are the control increments, $\hat{y}_{free}(t+j \mid t)$ is a j-step ahead prediction of the free response on data up to time t; $w(t+j)$ is a future set point or reference sequence and $\Delta\hat{y}_{forced}(t+i \mid t)$ is a j-step ahead prediction of the forced response on data up to time t. The output of the plant is used to update the nonlinear model state vector.

The incremental linear model has been obtained by identification and corresponds to a first order model with a delay of one.

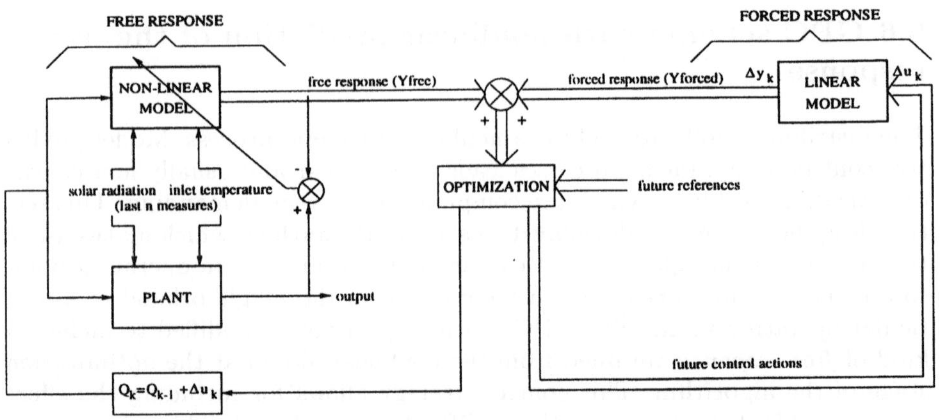

Fig. 5.28. Nonlinear prediction control scheme

5.6.2 Incremental formulation of predicted disturbances

In this section, a brief formulation of the GPC which incorporates the effect of measurable disturbances is presented. The significance of an incremental formulation of the prediction of measurable disturbances is discussed.

If a CARIMA model of the plant is used, taking into account the measurable disturbance term $v(t)$, the process can be described (the delay is considered as a part of $B(z^{-1})$ polynomial) by:

$$A(z^{-1})y(t) = B(z^{-1})u(t-1) + D(z^{-1})v(t) + C(z^{-1})\frac{\epsilon(t)}{\Delta}$$

which if multiplied by Δ takes the form:

$$A(z^{-1}) \Delta y(t) = B(z^{-1}) \Delta u(t-1) + D(z^{-1}) \Delta v(t-1) + C(z^{-1})\epsilon(t) \quad (5.54)$$

If $C(z^{-1}) = E_j(z^{-1})A(z^{-1}) \Delta + z^{-1}F_j(z^{-1})$ is introduced in the expression of the system model, the following prediction can be obtained:

$$\hat{y}(t+j) = \frac{F_j(z^{-1})}{C(z^{-1})}y(t) + \frac{E_j(z^{-1})B(z^{-1})}{C(z^{-1})} \Delta u(t+j-1) +$$

$$\frac{E_j(z^{-1})D(z^{-1})}{C(z^{-1})} \Delta v(t+j-1) + E_j(z^{-1})\epsilon(t+j)$$

If $E_j(z^{-1})B(z^{-1}) = G_j(z^{-1})C(z^{-1}) + z^{-j}\Gamma(z^{-1})$ and $E_j(z^{-1})D(z^{-1}) = P_j(z^{-1})C(z^{-1}) + z^{-j}\Upsilon(z^{-1})$, the optimal prediction of the system output can be obtained:

$$\hat{y}(t+j|t) = \frac{F_j(z^{-1})}{C(z^{-1})}y(t) + G_j(z^{-1}) \Delta u(t+j-1) + \frac{\Gamma_j(z^{-1})}{C(z^{-1})} \Delta u(t-1)$$

$$+P_j(z^{-1}) \triangle v(t+j-1) + \frac{\Upsilon_j(z^{-1})}{C(z^{-1})} \triangle v(t-1)$$

Which can be expressed by:

$$\mathbf{y} = \mathbf{G}\mathbf{u} + \mathbf{f}$$

where $\mathbf{f}$ is the system free response,

$$\mathbf{f} = \frac{F_j(z^{-1})}{C(z^{-1})} y(t) + \frac{\Gamma_j(z^{-1})}{C)(z^{-1})} \Delta u(t-1) + P_j(z^{-1}) \Delta v(t+j-1) + \frac{\Upsilon_j(z^{-1})}{C(z^{-1})} \Delta v(t-1)$$

As can be observed, the free response is composed of the following terms:

$\frac{F_j(z^{-1})}{C(z^{-1})} y(t)$: Effect of past outputs.
$\frac{\Gamma_j(z^{-1})}{C(z^{-1})} \triangle u(t-1)$: Effect of past control actions.
$\frac{\Upsilon_j(z^{-1})}{C(z^{-1})} \triangle v(t-1)$: Effect of past disturbances.
$P_j(z^{-1}) \triangle v(t+j-1)$: Effect over the future system output of the future disturbances. This term can be considered as an uncontrollable *forced response* included in the free response. In some cases the future disturbances (or load changes) are known. If they are not known they have to be predicted.

The GPC solution can be expressed as:

$$\mathbf{u} = (\mathbf{G}^T\mathbf{G} + \lambda \mathbf{I})^{-1} \mathbf{G}^T (\mathbf{w} - \mathbf{f})$$

If the first row of matrix $(\mathbf{G}^T\mathbf{G} + \lambda \mathbf{I})^{-1} \mathbf{G}^T$ is called $(\alpha_1 \quad \alpha_2 \quad \dots \quad \alpha_N)$, the control law can be expressed:

$$\Delta u(t) = -\sum_{i=1}^{N} \alpha_i \frac{\Gamma_i(z^{-1})}{C(z^{-1})} \Delta u(t-1) - \sum_{i=1}^{N} \alpha_i \frac{F_i(z^{-1})}{C(z^{-1})} y(t) - \sum_{i=1}^{N} \alpha_i \frac{\Upsilon_i(z^{-1})}{C(z^{-1})} \Delta v(t-1)$$

$$-\sum_{j=1}^{N} \sum_{i=1}^{N} \alpha_i \frac{P_{ij}}{C(z^{-1})} \triangle \hat{v}(t+j-1) + \sum_{i=1}^{N} \alpha_i w(t+i)$$

If

$R(z^{-1}) = C(z^{-1}) + \sum_{i=1}^{N} \alpha_i \Gamma_i(z^{-1}) z^{-1}$
$T_1(z^{-1}) = \sum_{i=1}^{N} \alpha_i z^{-N+i}$
$S(z^{-1}) = \sum_{i=1}^{N} \alpha_i F_i(z^{-1})$
$S_1(z^{-1}) = \sum_{i=1}^{N} \alpha_i \Upsilon_i(z^{-1})$
$S_2 = \sum_{j=1}^{N} \sum_{i=1}^{N} \alpha_i P_{ij}$

then:

$$R(z^{-1})\triangle u(t) = C(z^{-1})\,T_1(z^{-1})\,w(t+N) - S(z^{-1})y(t) - S_1(z^{-1})\triangle v(t-1)$$
$$-S_2\,C(z^{-1})\ \triangle v(t+j-1)$$

If the controller expression if multiplied by $z^{-1}B(z^{-1})$ and equation (5.54) by $R(z^{-1})$, and substituting the term that multiplies $\Delta u(t)$, the following expression can be obtained:

$$(\Delta A(z^{-1})\,R(z^{-1})+B(z^{-1})\,S(z^{-1})\,z^{-1})y(t) = B(z^{-1})\,C(z^{-1})\,T_1(z^{-1})\,w(t+N-1)$$
$$+C(z^{-1})\,R(z^{-1})\,\epsilon(t) - B(z^{-1})\,S_1(z^{-1})\,\Delta v(t-2) - B(z^{-1})\,S_2\,C(z^{-1})\,\Delta v(t+j-2)$$
$$+R(z^{-1})\,D(z^{-1})\,\triangle v(t)z^{-1}$$

In steady state conditions, all the terms accompanied by operator $\triangle$ are annulled, and $y(t) = w(t)$ for $t \to \infty$.

The conclusion which can be obtained from the previous expressions is the importance of using the disturbance term in an incremental form, to avoid steady state errors. All the expressions have been obtained supposing that a linear model relating the system output to the disturbances can be obtained.

The $C(z^{-1})$ polynomial can be used as a design filter. In this way, the optimal character of the predictions is lost, but good robustness properties can be achieved.

Using Ljung's notation [74], the model can be expressed in function of $M(z^{-1}) = B(z^{-1})/A(z^{-1})$, $N(z^{-1}) = D(z^{-1})/A(z^{-1})$ y $H(z^{-1}) = C(z^{-1})/\triangle A(z^{-1})$, obtaining:

$$y_m(t) = M(z^{-1})u(t) + N(z^{-1})v(t) + H(z^{-1})\epsilon(t)$$

where $M(z^{-1}), N(z^{-1})$ y $H(z^{-1})$ (with $H(0) = 1$) are rational functions in the z^{-1} operator. The exact representation of the plant is supposed to be:

$$y(t) = M_p(z^{-1})u(t) + N_p(z^{-1})v(t) + x(t)$$

where $x(t)$ is a general disturbance. The function $H(z^{-1})$ can be divided into $H(z^{-1}) = H_j^*(z^{-1}) + z^{-j}H_j(z^{-1})$ where $H_j^*(z^{-1})$ is a polynomial of order $j-1$. By manipulating these expressions, the predictor formulation can be obtained:

$$\hat{y}_m(t+j) = \frac{H_j^*(z^{-1})M(z^{-1})}{H(z^{-1})}u(t+j) + \frac{H_j^*(z^{-1})N(z^{-1})}{H(z^{-1})}v(t+j) + \frac{H_j(z^{-1})}{H(z^{-1})}y(t)$$

As:

$$\frac{H_j(z^{-1})}{H(z^{-1})}y(t) = y(t+j) - \frac{H_j^*(z^{-1})}{H(z^{-1})}y(t+j) =$$

$$= M_p(z^{-1})u(t+j) - \frac{M_p(z^{-1})H_j^*(z^{-1})}{H(z^{-1})}u(t+j)$$

$$-\frac{N_p(z^{-1})H_j^*(z^{-1})}{H(z^{-1})}v(t+j) + x(t+j) + N_p(z^{-1})u(t+j) - \frac{H_j^*(z^{-1})}{H(z^{-1})}x(t+j)$$

$$\hat{y}_m(t+j) = \frac{H_j^*(z^{-1})M(z^{-1})}{H(z^{-1})}u(t+j) + \frac{H_j^*(z^{-1})N(z^{-1})}{H(z^{-1})}v(t+j)$$
$$-\frac{H_j^*(z^{-1})M_p(z^{-1})}{H(z^{-1})}u(t+j) - \frac{H_j^*(z^{-1})N_p(z^{-1})}{H(z^{-1})}v(t+j)$$
$$-\frac{H_j^*(z^{-1})}{H(z^{-1})}x(t+j) + y(t+j)$$

$$\hat{y}_m(t+j) = y(t+j) + \frac{H_j^*(z^{-1})\check{M}(z^{-1})}{H(z^{-1})}u(t+j) + \frac{H_j^*(z^{-1})N(z^{-1})}{H(z^{-1})}v(t+j)$$
$$-\frac{H_j^*(z^{-1})N_p(z^{-1})}{H(z^{-1})}v(t+j) - \frac{H_j^*(z^{-1})}{H(z^{-1})}x(t+j)$$

with $\check{M}(z^{-1}) = M(z^{-1}) - M_p(z^{-1})$. The terms of this expression can be interpreted as:

- The effect of modeling errors between the control signal and the system output (unmodelled dynamics, $\frac{H_j^*(z^{-1})\check{M}(z^{-1})}{H(z^{-1})}u(t+j)$),
- the effect of modeling errors between the disturbances and the system output plus prediction errors of the disturbances ($\frac{H_j^*(z^{-1})N(z^{-1})}{H(z^{-1})}v(t+j) - \frac{H_j^*(z^{-1})N_p(z^{-1})}{H(z^{-1})}v(t+j)$)
- and the effect of disturbances ($\frac{H_j^*(z^{-1})}{H(z^{-1})}x(t+j)$).

If $j = 1$ is chosen, $H_1^* = 1$, the transfer function of the filters can be expressed as $\Delta A/C$ which corresponds to bandpass filters. The Δ term eliminates prediction errors due to errors in static gain estimation in models $M(z^{-1})$ and $N(z^{-1})$. Constant value disturbances are also eliminated. If the order of $C(z^{-1})$ is high, the component of prediction errors due to unmodelled dynamics, which is great at high frequencies, is also eliminated.

In cases with great modeling errors, it is convenient to use GPC with filtered signals obtained from the input/output data of the plant in the following way:

$$y^f(t) = \frac{\Delta}{C(z^{-1})}y(t);\ u^f(t) = \frac{\Delta}{C(z^{-1})}u(t);\ v^f(t) = \frac{\Delta}{C(z^{-1})}v(t);$$

As has been mentioned before, an important conclusion is the convenience of using a filtered disturbance prediction in an incremental form to avoid steady state errors.

The free response can be obtained (supposing null future noises and disturbances):

$$\hat{y}_m(t+j)|_{free} = y(t+j)|_{free} + \frac{H_j^*(z^{-1})}{C(z^{-1})} A(z^{-1})\Delta(N(z^{-1})v(t+j) - N_p(z^{-1})v(t+j))$$

In this expression, the importance of the term $\Delta/C(z^{-1})$ in the achievement of an adequate free response can be analyzed. This term represents a bandpass filter that eliminates the continuous component and high frequency components of the future disturbances prediction.

5.6.3 Application to the distributed solar collector field

As has been indicated, the idea described in this section is based on separating the system response obtained with GPC control into *free response* and *forced response*. Free response is calculated supposing that at the next sampling period the control signal will remain constant. By using an adequate nonlinear model and from a prediction of the measurable loads (mainly inlet temperature and inlet solar radiation), this *free response* can be calculated. The *forced response* is calculated by applying the GPC algorithm to an adequate linear model.

Nonlinear prediction of the free response

The nonlinear prediction model of the free response of the plant has been obtained using the same equations as describe the nonlinear simulation model analyzed in chapter 2, but increasing the length of all the integration segments to the maximum length allowed by the approximation used for the numerical integration of said equations. Figure 5.29 shows a comparison between the original nonlinear model response and that of the simplified model.

Given that the sampling time is high (39 seconds), the calculation time is sufficient for adopting a nonlinear description of the system of the type indicated. An added problem to that of calculating a nonlinear model which accurately reflects the plant behaviour throughout the entire range of work is to have a good load prediction model.

Prediction of the inlet oil temperature

Inlet oil temperature variations are difficult to predict because the temperature profile in the tank can be variable and the instant at which oil from different thermoclines leaves the tank depends on the oil flow. Another reason for the inlet oil temperature variations is the recirculation of the hot oil using the three-way valve till it reaches the temperature adequate for entering the top of the storage tank.

Some studies have been performed [26] as shown in chapter 3 to relate changes in the outlet oil temperature to the inlet oil temperature and solar radiation variations. The benefits provided if a dynamical model is used to account for inlet oil temperature variations are not considered to justify the added complexity or uncertainty involved. Thus, for prediction purposes, the

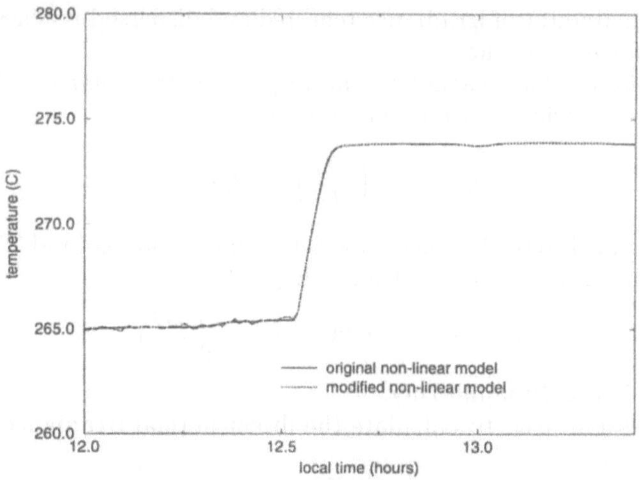

Fig. 5.29. Response with nonlinear model and simplified model

inlet oil temperature has been considered constant at the prediction horizon in the optimization part of the algorithm.

Solar radiation prediction model

Solar radiation undergoes changes due to its daily cycle and to passing clouds. The prediction of passing clouds is very complex, and so, an approximation based on operating on a clear day has been adopted.

There are many models of solar radiation evolution, where the solar radiation profile depends on factors such as geographic localization, the day of the year, etc. Here, the approximation given by Castro [30] has been adopted.

The magnitude of the solar radiation at a determined point on the earth surface varies mainly with the geographical localization and the day of the year, and also with the meteorological characteristics of the chosen instant.

To calculate the global irradiance level on a surface, the extraterrestrial irradiance level I_c (solar constant[2]) must be known. This value reaches the atmosphere of the Earth with a small modification I_{ext}, due to the elliptical trajectory of the Earth around the Sun.

As the radiation reaches the atmosphere, fragmentation is produced, one part is reflected outside, and another part is absorbed by the atmosphere; the rest penetrates the atmosphere. One part of this is dispersed, the other is reflected and the rest reaches the Earth in form of direct radiation I_b.

[2] This is the amount of total energy that contains the extraterrestrial solar irradiance, integrated in all the spectrum of wave lengths. The value used is $I_c = 1367$ W/m^2.

Part of the amount of irradiance reflected and dispersed reaches the Earth in form of diffuse irradiance.

The irradiance that reaches the atmosphere is the solar constant multiplied by the eccentricity correction factor E_{xc}:

$$I_{ext} = I_c \left(\frac{D_0}{D}\right)^2 = I_c E_{xc}$$

D being the Sun-Earth distance at a determined instance and D_0 its mean value. The chosen formula to calculate E_{xc} is:

$$E_{xc} = 1 + 0.033 \cdot cos(\frac{2 \cdot \pi \cdot JD}{365})$$

in which JD is the Julianne Day.

The expression used to calculate the direct normal irradiance for a clear day (I_{bn}) [30] is:

$$I_{bn} = 0.9751 \cdot I_{ext} \cdot T_b \qquad (5.55)$$

where T_b is the transmittance of the direct normal irradiance, which depends on factors such as the ozone cover thickness, the thickness of the water steam cover, two coefficients of atmospheric cloudiness and the air mass at standard pressure.

Expression 5.55 has been used to predict the evolution of the solar radiation at a predetermined prediction horizon. Some examples of the result obtained when using the clear day solar radiation prediction model are presented and compared to the actual direct solar radiation measured at the plant (Fig. 5.30, 5.31 and 5.32).

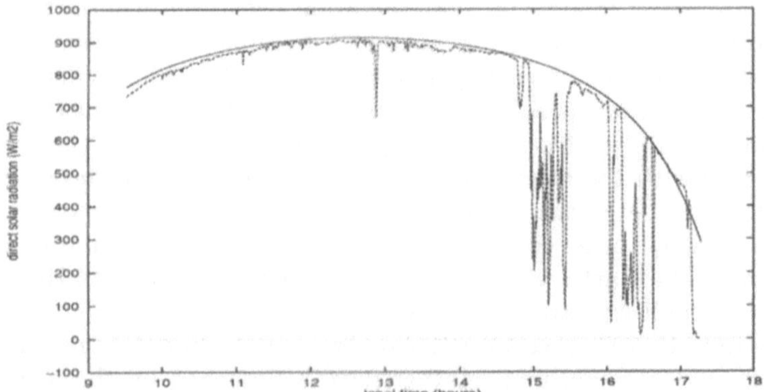

Fig. 5.30. Direct solar radiation and clear day prediction (26/11/90)

- The first plot (26/11/90) corresponds to a day with high solar radiation levels in which, at the end of the daily operation, large clouds covered the field.

- The second plot (19/09/91) corresponds to a covered day with low solar radiation levels, but without the presence of passing clouds.
- The third plot (02/03/93) corresponds to a covered day with passing clouds.

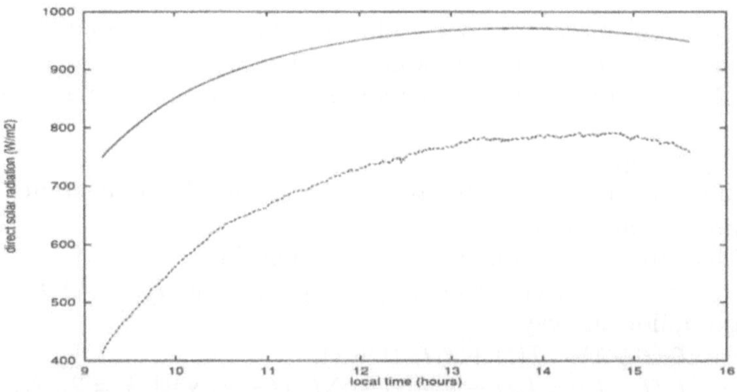

Fig. 5.31. Direct solar radiation and clear day prediction (19/09/91)

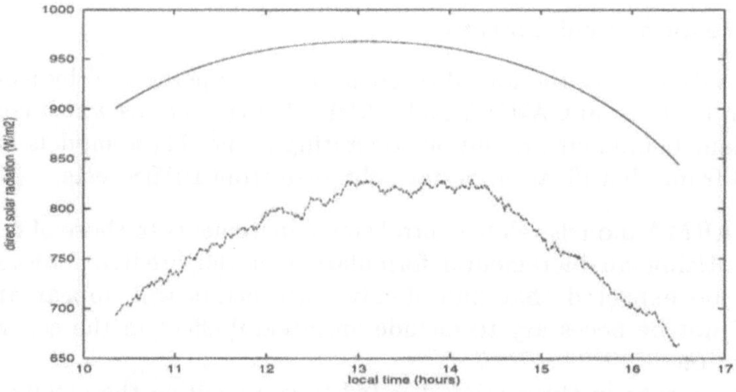

Fig. 5.32. Direct solar radiation and clear day prediction (02/03/93)

As can be seen, lines have been drawn depending on the local hour. By using the prediction model, a filtered value of the solar radiation evolution is directly obtained. On the other hand, the model is valid not only for clear days but also for covered days without passing clouds, because in such cases, there is an almost constant difference between the measured values and the predicted ones, and an incremental formulation of the model has been chosen.

In the case of small prediction horizons, a constant value equal to the actual one can be used in the predictions. Also predictions based on extrapolation of previous values have been evaluated, but worse results have been obtained.

The completely free response prediction scheme consists mainly of the following steps:

1. Calculate off-line the clear day direct solar radiation prediction. Values obtained from this calculation are called $I_m(t)$.
2. Calculate solar radiation prediction increments $\Delta I_m(t) = I_m(t) - I_m(t-1)$.
3. During the operation:
   a) Measure the inlet oil temperature and consider it constant along the prediction horizon.
   b) Measure the real direct solar radiation $I(t)$.
   c) Perform the direct solar radiation predictions $I_p(t+k), k = 1...N$ in the following way:
      i. $I_p(t+1) = I(t) + \Delta I_m(t+1)$.
      ii. $I_p(t+k) = I_p(t+k-1) + \Delta I_m(t+k)$ with $k = 2...N$.

      In this way, an implicit feedback of the disturbance values is obtained.
   d) Use the predicted values in the nonlinear model to calculate the free response (calculated considering $u(t) = u(t+1) = ... = u(t+N-1)$).

Forced response calculation

The evaluation of the forced response has been performed by using linear models in both usual CARMA and CARIMA formulations which correspond to the plant behaviour around one operating point. These models have been obtained from identification of data obtained from PRBS tests.

- The CARMA models relate control signal increments to those of the output signal. Using an incremental formulation of the predicted disturbances, it can be expected that null steady state errors will appear and so, it should not be necessary to include an integral effect in the control signal calculation.
- As can be seen in the section devoted to commenting the results obtained at the plant, there are some factors which can produce steady state errors, such as the consideration of values of reflectivity different to the real one in the nonlinear model.

One advantage of using the nonlinear model of the plant in the free response calculation is the ability to take into account the dynamical effect of the disturbances (feedforward control), producing the input signal corresponding to the actual operating point (this model accounts for changes in variables different from the oil flow). If set point changes are small (less than

20°C), the hypothesis of operation around an operating point in the forced response calculation is a right one. The value of the model polynomials are:

$$\begin{aligned} A(z^{-1}) &= 1 - 0.89973z^{-1} \\ B(z^{-1}) &= -0.629035z^{-2} \end{aligned}$$

The delay has been included in the $B(z^{-1})$ polynomial. This model has been obtained for the operation about medium flows. Model polynomials have to be changed to take into account other different operating points. In this particular case, due to the strong influence of solar radiation on the absolute value of the control signal, the contribution of the linear part in the control signal is smaller than that of the free response. In this way, even in cases of erroneous estimation of A and B polynomials, good results can be achieved. As can be seen, the model has negative steady state gain, because of the relation between oil flow changes and outlet temperature variations.

As has been mentioned, the incremental oil flow signal can be obtained both with a CARMA and CARIMA formulation. In the latter case, the control signal can be obtained by minimizing the GPC cost function, obtaining well-known expressions:

$$\Delta u(t) = (\mathbf{G}^T\mathbf{G} + \lambda\mathbf{I})^{-1}\mathbf{G}^T(\hat{\mathbf{y}}_{free}(t) - \mathbf{w}(t)), \quad u(t) = u(t-1) + \Delta u(t).$$

5.6.4 Simulation studies

As in previous studies, a series of simulations to choose the adequate load prediction models were carried out. The nonlinear model used for free response calculation, obviously, works well in this case, because as a plant it uses the complete simulator which, as has been mentioned, behaves very similarly to the one used for calculating the free response. In order to analyze the effect of predicting the solar radiation, a trial has first been studied in which the inlet temperature scarcely changes. Figure 5.33 shows the evolution of the direct solar radiation (clear day). A control and prediction horizon has been chosen of value 10 and a value of $\lambda = 50$ from simulations.

The direct solar radiation is, in this case, predicted along the chosen horizon using two different prediction models:

1. One consisting of the straight line that best adjust itself to the solar radiation in the last 10 sampling periods.
2. Another with the incremental formulation using the direct solar radiation predicted for the clear day model.

As can be seen in the illustrations (Figures 5.34 and 5.35) the temperature follows the reference with a very short response time at all set point changes. This is due to the fact that the free response is calculated from a nonlinear model which takes into account the changes in the process dynamics which

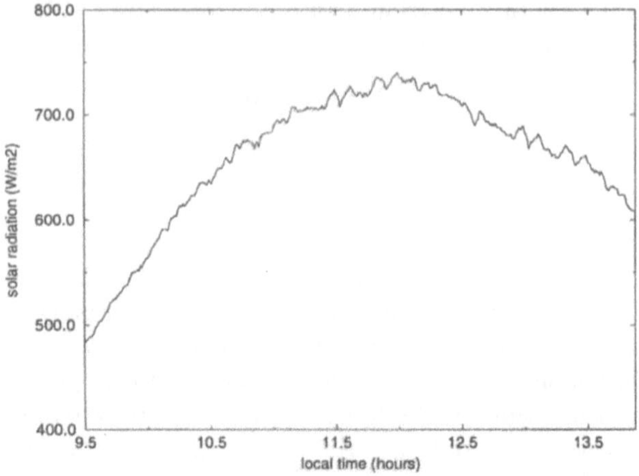

evolution of solar radiation (W/m^2)

Fig. 5.33. Solar radiation conditions for simulations

take place when the operating conditions are changed. If the linear structure of the control is formulated using an incremental model, a good prediction of the free response is needed in order to avoid errors in the permanent regime at the outlet temperature.

Figure 5.34 presents the results obtained with a prediction model of direct solar radiation based on a straight line prediction (that which better adjusts to the last N measurements). The results are good, but with this type of prediction, if there is a long horizon, great deviations appear in the predictions which give rise to the oscillations that appear in the response. Figure 5.35 shows the comparison between the outlet and that obtained using the incremental prediction model of direct solar radiation on a clear day. Better results are obtained in this last case. Obviously, in the case of passing clouds, no model is capable of predicting the future direct solar radiation.

Regarding the prediction of the inlet temperature, it can be seen that the consideration made (it remains constant at the prediction horizon) produces very good results in the proposed control scheme. Figure 5.36 shows the evolution of the outlet temperature (dashed-line) when faced with great changes in the inlet temperature (dotted line). It can be seen that the control scheme reacts adequately to such changes, the outlet temperature quickly reaching the reference value after the disturbances.

Many other simulations have been carried out (results are not included here because of the lack of space). In general the results obtained are quite satisfactory on smogy days. In the case of passing clouds the behaviour is not so satisfactory due to the fact that the radiation prediction model always presupposes positive increases before the solar midday and negative ones afterwards.

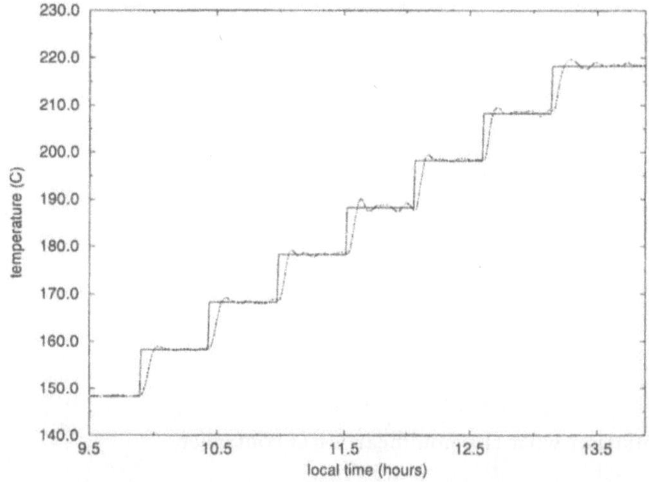

evolution of the reference and outlet temperatures (^{o}C)

Fig. 5.34. Simulation with the nonlinear GPC controller

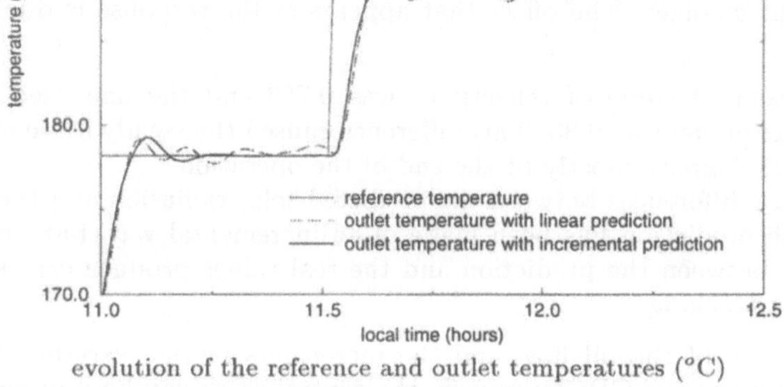

evolution of the reference and outlet temperatures (^{o}C)

Fig. 5.35. Simulation with the nonlinear GPC controller

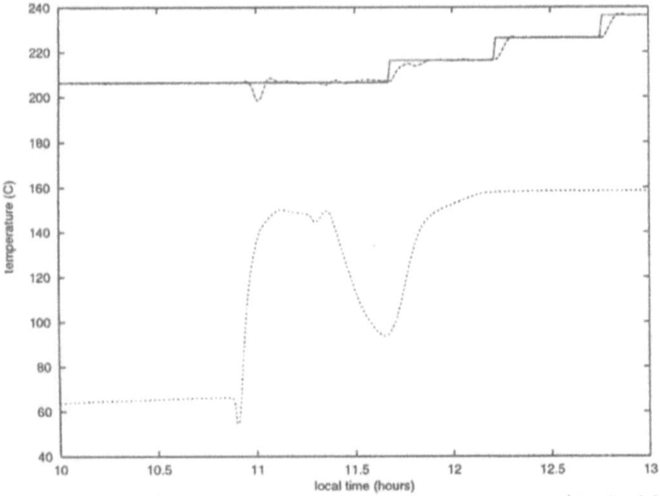

inlet temperature (dotted line) and output temperature (dashed line)

Fig. 5.36. Simulation with the nonlinear GPC controller

5.6.5 Plant results

The control algorithm presented in this section has been applied to control the distributed solar collector field ACUREX. Two representative tests have been chosen to show the main results achieved with this approach. In the first test, a CARMA model has been used to obtain the forced response of the plant. In the second test, a CARIMA formulation has been used.

Figure 5.37 shows the evolution of the characteristic variables of the test using the nonlinear model of the field with incremental prediction of the direct solar radiation and a CARMA model to obtain the forced response. As can be seen, after an initial transient due to large passing clouds, the outlet temperature reaches the set point. The controller produces a fast response to set point changes. The offset that appears in the response is due to two factors:

- The measured value of reflectivity was 0.753 and the one used by the nonlinear model was 0.86. This difference caused the steady state errors of about 1.5 degrees, mostly at the end of the operation.
- There are differences between the predicted solar radiation and the actual one. The prediction has been made in an incremental way, but great differences between the prediction and the real values produce errors in the set point tracking.

The evolution of the oil flow and the increments in this variable are also shown. By using the GPC approach, the control effort can be diminished by increasing the value of the cost factor λ to avoid sudden changes in the oil flow impelled by the pump to assure a long life cycle of this actuator.

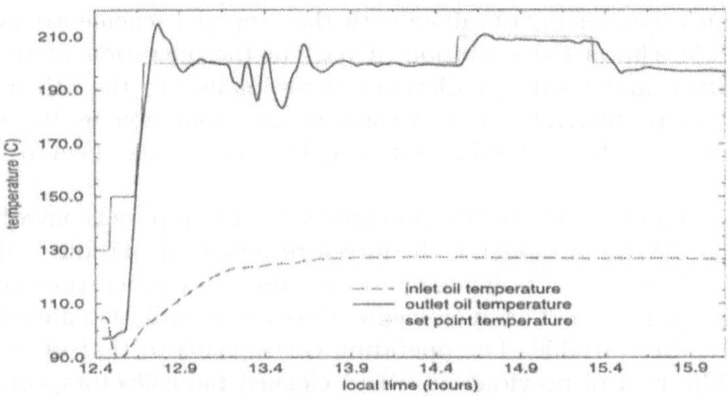

set point, inlet and outlet temperatures (°C)

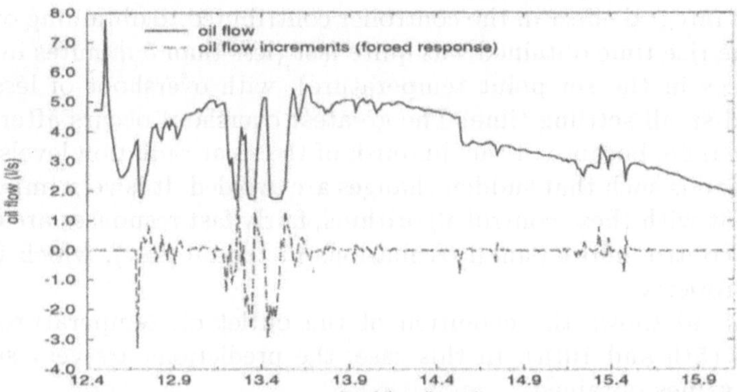

oil flow and increments (l/s)

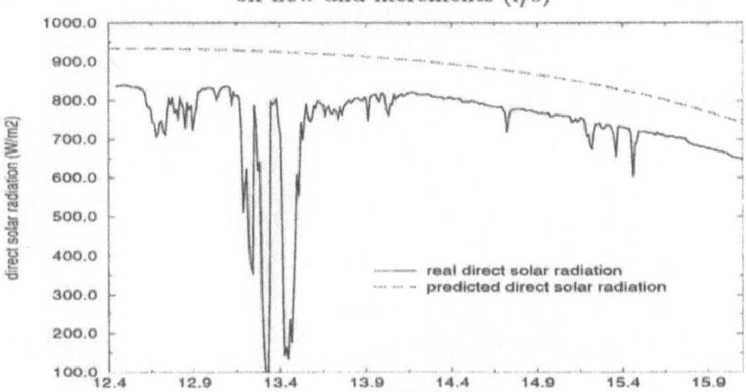

direct solar radiation and prediction (W/m²)

Fig. 5.37. Test with the nonlinear GPC controller (30/10/95)

The evolution of the direct solar radiation corresponds to a day with large passing clouds and smogy atmosphere. In this case, the prediction differs from the real values, but results obtained with the proposed scheme are good.

Figure 5.38 shows the evolution of part of the operation of the outlet oil temperature and future predictions corresponding to the 5th and 10th sampling time respectively. In this case, future prediction in 10 sampling times differs from the real value obtained because of the effect of passing clouds.

Figure 5.39 presents the results obtained with the proposed control scheme considering a CARIMA model in the forced response calculation. This figure shows the evolution of set point temperature, inlet oil temperature, outlet oil temperature, oil flow, oil flow increments, direct solar radiation and clear day prediction of this variable. The operation corresponds to that of an almost clear day. The rain of previous days had cleaned the collectors, originating high values of reflectivity (not known when the test was performed). The inclusion of integral effect in the controller contributed to obtaining offset-free control. The rise time obtained was quite fast (less than 5 minutes in changes of 10 degrees in the set point temperature), with overshoot of less than 3 degrees and small settling time. The greatest overshoot occurs after the last set point change, because of the diminish of the solar radiation levels. The oil flow evolution is such that sudden changes are avoided. It is very important to mention that with these control algorithms, fairly fast responses are obtained with little control effort (small oscillations in the oil flow), which is a very desirable property.

Figure 5.40 shows the evolution of the outlet oil temperature and its predictions (5th and 10th). In this case, the predictions are very similar to the actual values obtained.

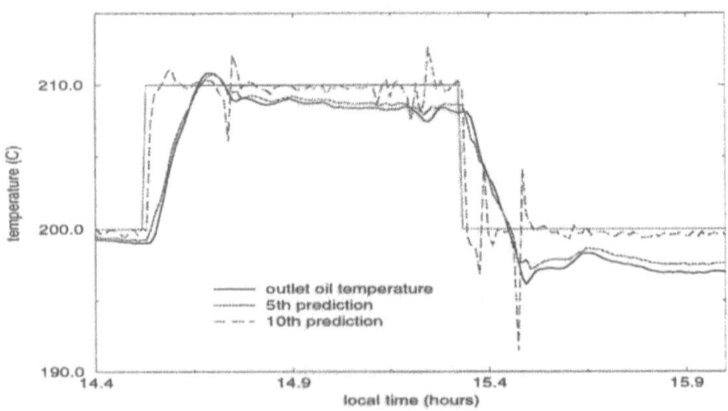

Fig. 5.38. Test with the nonlinear GPC controller (30/10/95)

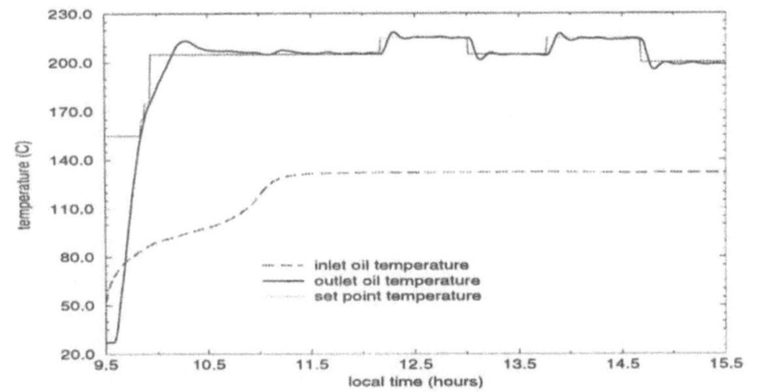

set point, inlet and outlet oil temperatures (°C)

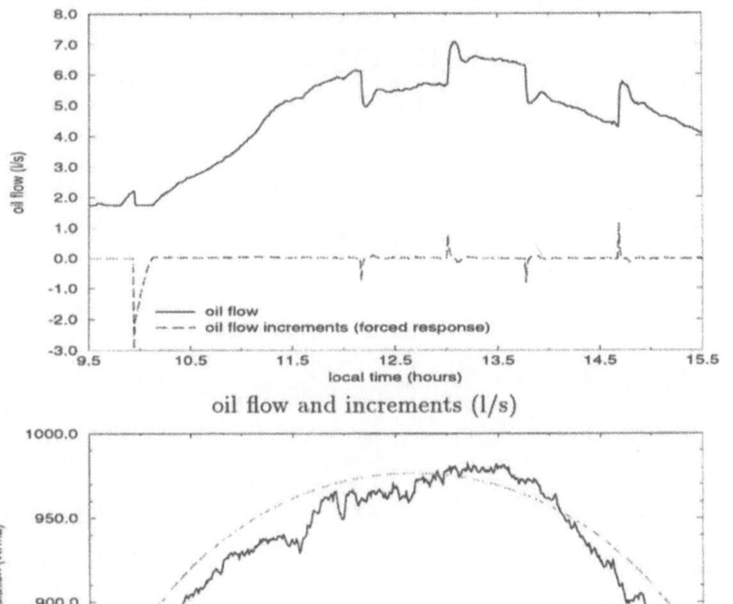

oil flow and increments (l/s)

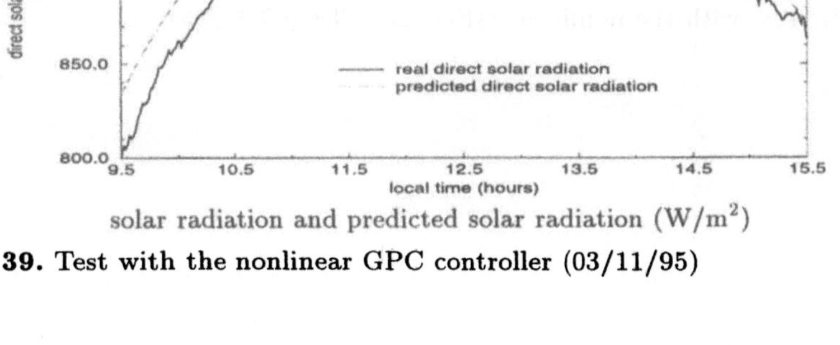

solar radiation and predicted solar radiation (W/m^2)

Fig. 5.39. Test with the nonlinear GPC controller (03/11/95)

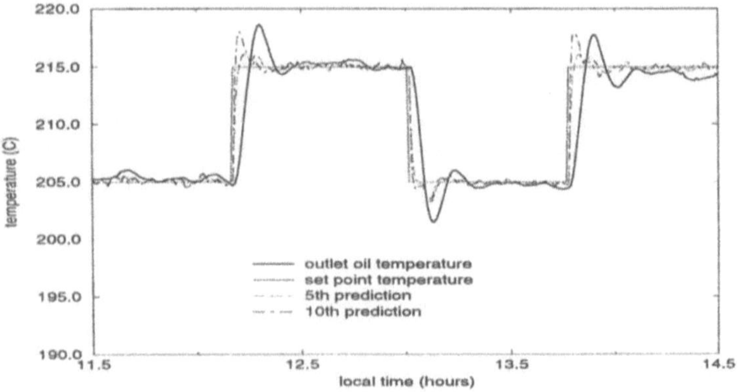

outlet temperature and predictions (°C)

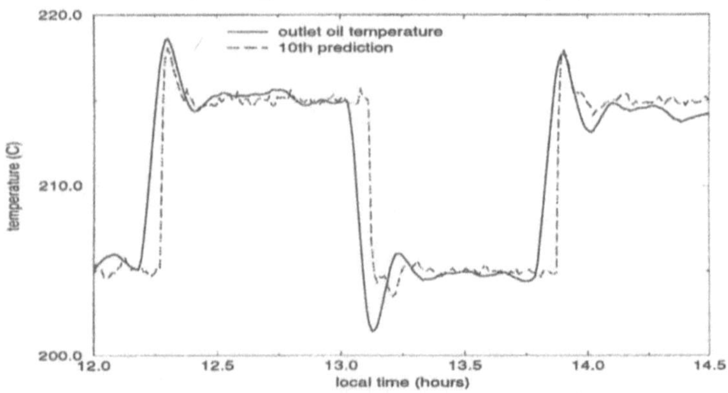

outlet oil temperature and prediction 10 (°C)

Fig. 5.40. Test with the nonlinear GPC controller (03/11/95)

Disturbance rejection

The nonlinear GPC controller has demonstrated to have good disturbance rejection properties. Figure 5.41 contains the result obtained when operating with large clouds (direct solar radiation variations of more than 700 W/m^2) and inlet oil temperature variations of about 10°C, which lead to oil flow saturations during a great amount of sampling times. After saturations, the control system reacts so that the set point is again reached with an overshoot of less than 8°C.

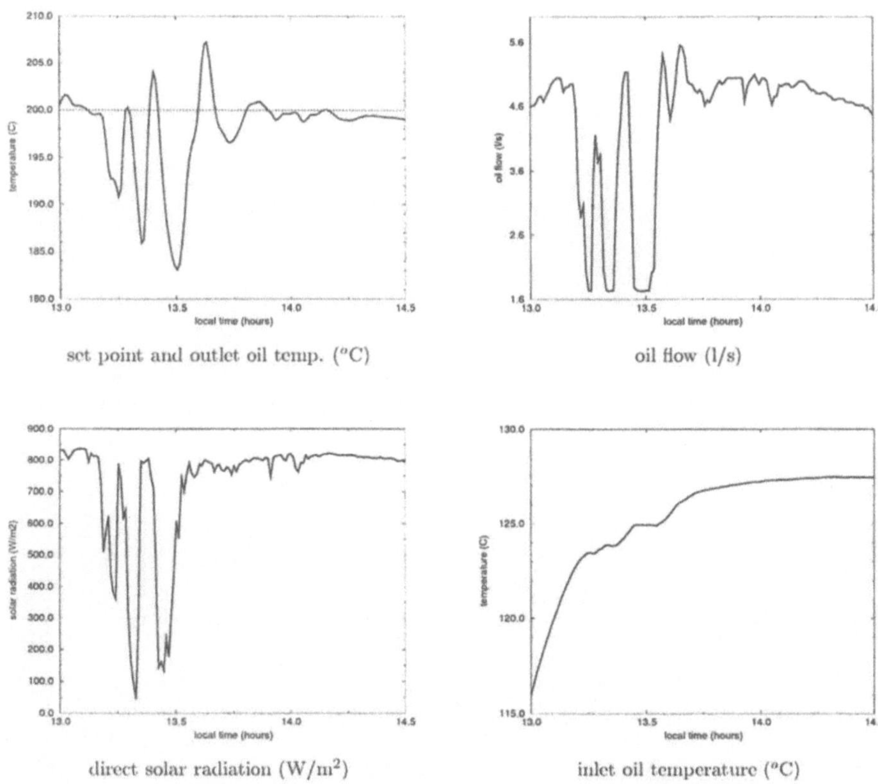

Fig. 5.41. Disturbance rejection with the nonlinear GPC controller (30/10/95)

CHAPTER 6
FREQUENCY DOMAIN CONTROL AND ROBUST OPTIMAL CONTROL

In this chapter, two control schema based on the frequency response of the open loop system are studied.

Firstly, an adaptive frequency-based internal model control (IMC) strategy is considered. The key idea of the method is to implement both the model of a plant and the controller with frequency-based interpolation models. The control scheme tries to identify a series of points on the Nyquist diagram of the system, and particularly, some near of the main antiresonance mode, so that the controller can take into account these characteristics of the plant in the frequency domain. The performance of the algorithm is first compared to classical controllers using linear models. The nonlinear computer model of the collector field is used to study the improvements that can be achieved when operating the plant using the proposed algorithm.

Secondly, the development of an optimal LQG/LTR controller is considered. Modern control methods have a different design philosophy to that of the classical frequency-based methods. Nevertheless, a connection between those methodologies is established in relation to the robustness of the controlled system. By using the loop transfer recovery mechanism (LTR) in combination with a normal optimal controller design, the stability margins of the controlled system (which are properties typically linked to the frequency domain) can be augmented when an observer is included in the control scheme. In this chapter, a frequency domain study of the LQG method [100] is introduced.

6.1 Adaptive frequency domain internal model control

6.1.1 Introduction

The design and application of an internal model adaptive controller using a frequency domain approach is considered in this section. The controller is designed generically to take into account systems with resonance and antiresonance modes, being able to be applied to the distributed solar collector field. The basic idea of the method is to implement both the plant and the controller using frequency domain interpolating models. The algorithm is first analyzed in a first stage of the design using a linear model of a system which simulates the effects of an antiresonance mode. Afterwards, using the nonlinear distributed parameter model, the advantages and drawbacks of the real time implementation of the algorithm are discussed. Finally, some results obtained when operating with the plant are explained [11], [12].

Adaptive control schema which use the certainty equivalence principle tend to be quite sensitive to the choice of the structure of the models to be estimated, poor robustness characteristics usually being obtained when using low order models. A possible solution to this problem is to use a frequency domain description of the plant, identifying the frequency response of the system in a series of points. As is well known, a drawback of the adaptive controller design methods (usually pole-placement) is the direct relation between parametric model order and controller complexity. If working with simple controllers is desirable, it is necessary to estimate low order models, which will model the system dynamic characteristics only in a narrow frequency band.

In the proposed method, the idea is to find a good process model in a sufficiently wide frequency range and then to use an approximate method to find low order controllers.

As has been mentioned in chapter 3, tuning methods of classical controllers (PID) are often based on time domain characteristics such as open loop step response methods. In many cases, good behaviour can be achieved. When an oscillatory response is obtained, controllers have to be detuned in order to increase the stability and robustness of the controlled system. Designers tend to interpret oscillatory behaviour as a consequence of wrong tuning of the controller, or errors in the identification and adaptation mechanism (in the adaptive case). Nevertheless, there are cases such as the one treated in this chapter (systems with antiresonance modes in the control bandwidth) in which the open loop step response does not show (or shows only small) antiresonance characteristics or oscillations, but when demands are made on the controlled system, *kickback* characteristics or oscillatory behaviour appear as a consequence of exciting the antiresonance modes.

The advantages of design methods based on a frequency domain analysis have been treated in chapter 1. In the field of adaptive control, and in relation to the method developed in this chapter, it is important to mention works of Hägglund and Aström, Middelton and Källén and Wittenmark included in the references [53], [83], [57] respectively, and mainly Tang, Ortega, Camacho and Flores [116], [118], [27], [117].

The adaptive algorithm presented in this chapter is based on frequency domain techniques that use an internal model control structure to control systems with antiresonance modes. Although it can easily be shown that the IMC structure is equivalent (if there are no constraints) to the standard unit feedback structure, IMC allows for a convenient parameterization of all stabilizing controllers for a stable plant. This property considerably simplifies the task of designing *stable* adaptive controllers. Another advantage of IMC in respect to the standard unit feedback configuration is that constraints can be handled more efficiently.

6.1.2 The IMC control structure

Internal model control is an alternative structure to the standard unit feedback structure. Its main advantage is based on the fact that, in the case of a perfect model, the closed loop is stable if a stable controller is selected. IMC combines both the advantages of open loop control, that is , great simplicity in the controller design, and that of the closed loop, which allows system model uncertainties and disturbances to be compensated for. Another fundamental advantage is the possibility of handling control signal constraints, due to the fact that the structure permits the injection of the saturated signal both to the plant and plant model. The basic control structure is shown in Figure 6.1.

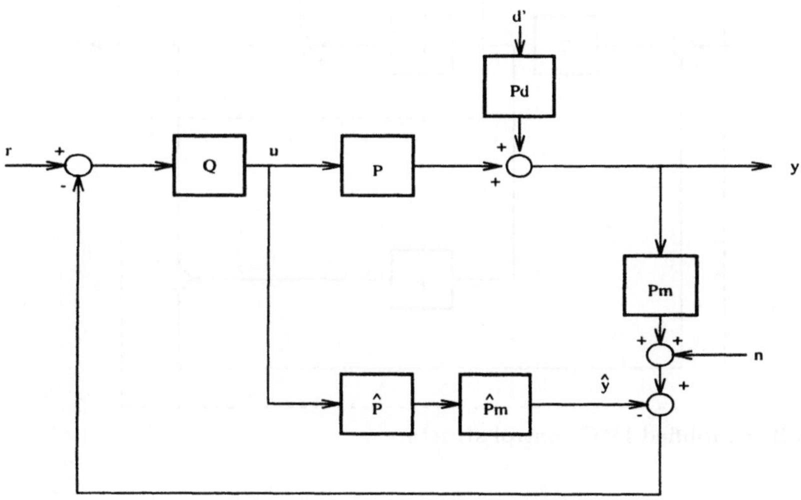

Fig. 6.1. IMC control structure

P: Represents the system transfer function, which is not usually known exactly.
P_m: Transfer function of the measurement device.
$\hat{P}$: Transfer function of the system model.
$\hat{P}_m$: Transfer function of the model of the measure device.
P_d: Transfer function which describes the effect of the disturbance on the output variable.
Q: Controller transfer function.
A noise n is added to the measure of y.

The objective of the control strategy is to maintain the output of the system as near as possible to the reference. A simplified diagram can also be used, as shown in Figure 6.2. In such a diagram, the output is supposed to be known exactly ($P_m = 1$, $n = 0$). Notice that the controller is constituted by the part of the block diagram inside the dotted-line box, which includes a model of the plant. This is the reason why this control structure is known as internal model control.

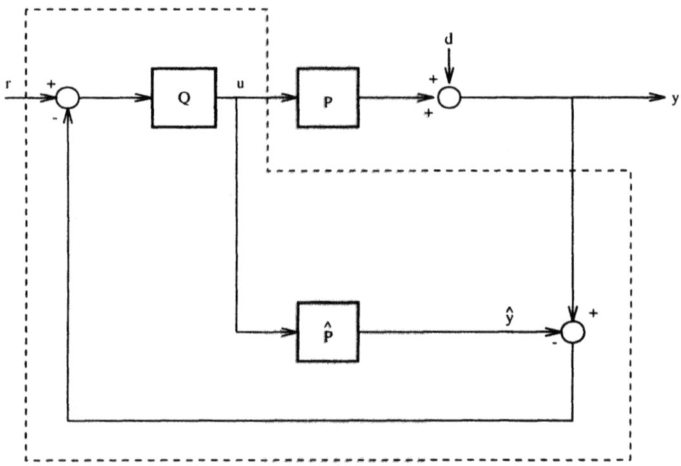

Fig. 6.2. Simplified IMC control structure

The feedback signal is $\hat{d} = (P - \hat{P})u + d$. If the model is exact ($P = \hat{P}$) and there are no disturbances ($d = 0$), the output of the model is equal to that of the system, and so the feedback signal is zero and an open loop control scheme is obtained. The feedback is included so as to account for system uncertainties and disturbances.

6.1.3 Stability and performance of IMC

Internal stability

To analyze the internal stability of the IMC control scheme, the transfer functions between all the possible inputs and outputs of the process have to be examined. As can be seen in Figure 6.3, three independent inputs to the system (r, u_1 and u_2) and three independent outputs (y, u and $\hat{y}$) exist. If there is no modelling error ($P = \hat{P}$), the following matrix can be constructed:

$$\begin{pmatrix} y \\ u \\ \hat{y} \end{pmatrix} = \begin{pmatrix} PQ & (1-PQ)P & P \\ Q & -PQ & 0 \\ PQ & -P^2Q & P \end{pmatrix} \begin{pmatrix} r \\ u_1 \\ u_2 \end{pmatrix} \tag{6.1}$$

It can easily be verified, by inspecting (6.1), that *If the model is perfect ($P = \hat{P}$),the IMC controller is internally stable if and only if both plant P and controller Q are stable* (Morari and Zafiriou, 1989 [86]).

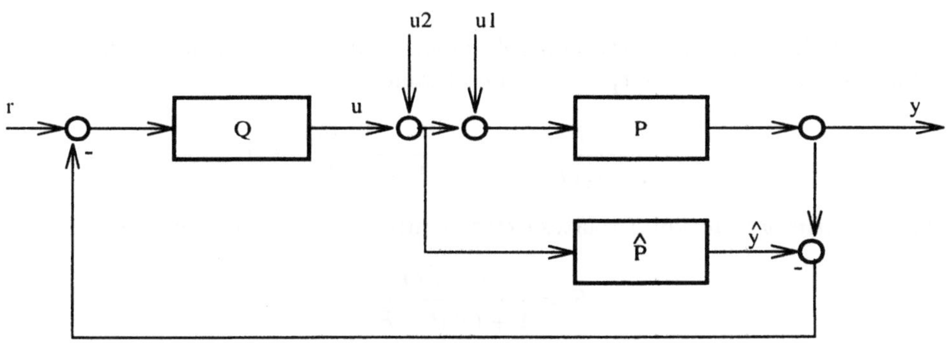

Fig. 6.3. Diagram to analyze the internal stability of IMC

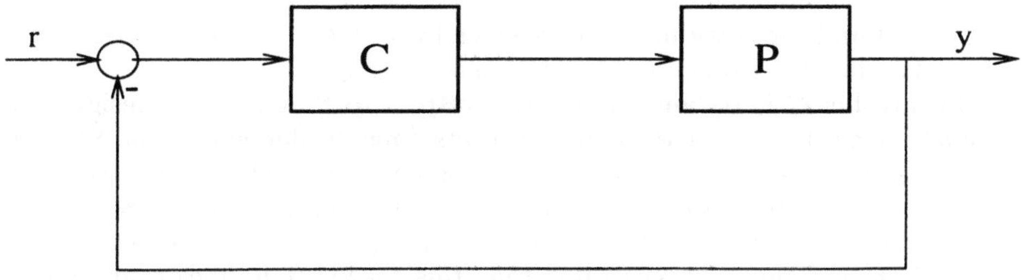

Fig. 6.4. Standard feedback controller

and $H_k(z)$ are bandpass filters with center frequency ω_k defined as:

$$H_k(z) = \frac{H_c(z)}{\alpha_k}\frac{1}{1-z^{-1}} \quad \text{for } k=0 \tag{6.10}$$
$$H_k(z) = \frac{H_c(z)}{\alpha_k^2+\beta_k^2}\frac{1}{1-2cos(\omega_k)z^{-1}+z^{-2}} \quad \text{for } k=1,2,\cdots,L$$

The interpolation model is composed of a bank of bandpass filters in cascade with first order systems with finite impulse response (FIR), as can be seen in Figure 6.5. By appropriately choosing coefficients c_k and d_k the system shown in Figure 6.5 will have an arbitrary frequency response at the interpolating point ω_k.

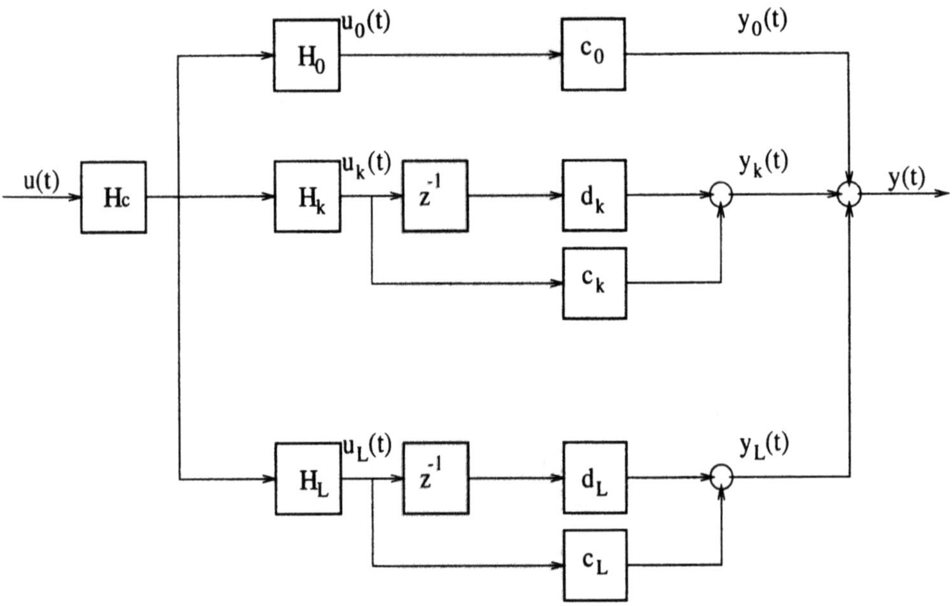

Fig. 6.5. Frequency interpolation

When coping with equally separated points, that is, $\omega_k = 2\pi k/N$ (Figure 6.6) it can be demonstrated that $\alpha_k = 1$, $\beta_k = 0$ for $k = 1, 2, \cdots, L$, and that the *comb* filter (Figure 6.7) is reduced to:

$$H_c(z) = \frac{1-z^{-N}}{N} \tag{6.11}$$

As there is a pole-zero cancelation on the unitary circle (6.3), in order to avoid filter instability, z must be substituted by z/ζ, ζ being a scalar slightly greater than one.

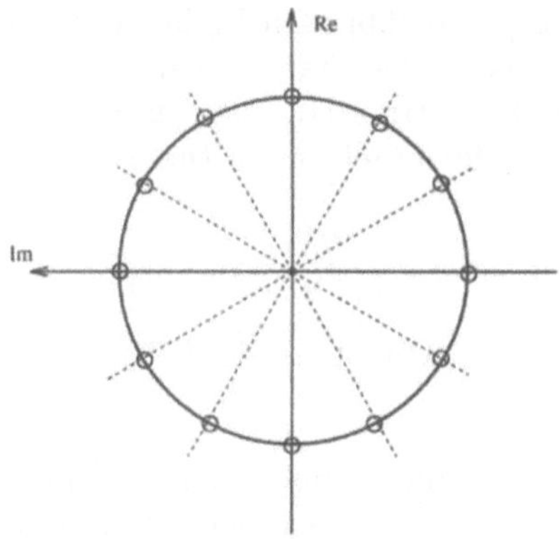

Fig. 6.6. Interpolating points location

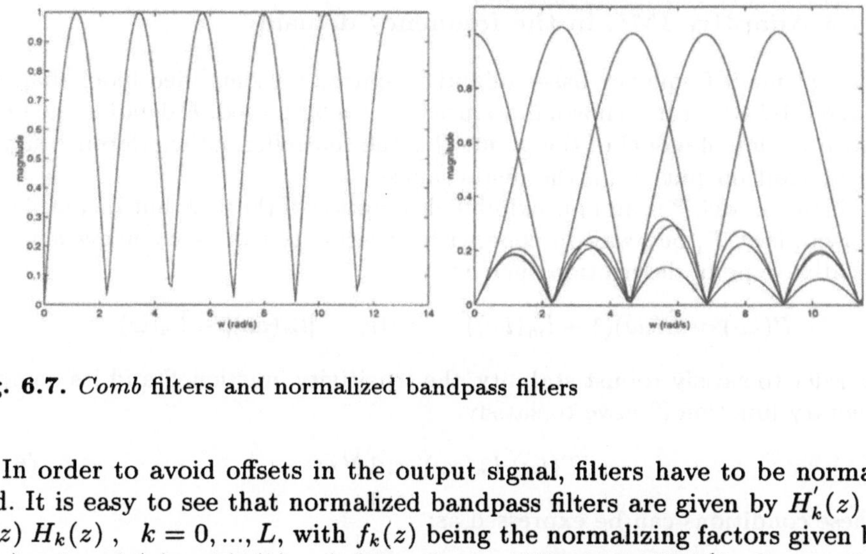

Fig. 6.7. *Comb* filters and normalized bandpass filters

In order to avoid offsets in the output signal, filters have to be normalized. It is easy to see that normalized bandpass filters are given by $H_k^{'}(z) = f_k(z)\, H_k(z)\ ,\ \ k = 0, ..., L$, with $f_k(z)$ being the normalizing factors given by $f_0(z) = 1$ y $f_k(z) = 2(\alpha(k) - (\alpha(k)cos(w_k) - \beta(k)sin(w_k))z^{-1})\ \ \ k = 1...L$.

The use of bandpass filters causes the data entering the estimators to be filtered and so, it is logical to think that a low order model can be identified.

The choice of width of the bandpass filters is subject to two contradictory facts:

- They must be chosen to be very narrow to adequately model the central frequency.
- They must be sufficiently wide to have a limited time response (not too high). Moreover, narrow filters have a long lasting impulse response and poles very close to the stability limit.

As indicated in [117] and [56], if a higher phase lag has to be included in some frequency bands, some changes can be performed in the transfer function which has to be estimated, such as including a z^{-m} term or by using other kinds of transfer functions in the estimation, such as:

$$b/(z-a)$$
$$b/(z^2+a_1z+a_2)$$
$$(b_0z+b_1)/(z+a_1)$$

These structures, or other similar ones, are convenient when the estimation of the frequency response is to be carried out near a resonance, where the magnitude and phase suffer great fluctuations. In the section devoted to commenting the results obtained in the operation, a comparison (Figure 6.21) of the results obtained with two different models is performed.

6.1.5 Adaptive IMC in the frequency domain

To implement frequency based adaptive control, the simplified block diagram of the IMC structure (shown in Figure 6.2) has been used. P denotes the plant and $\hat{P}$ a control model of the plant. Q is the controller, r the reference signal, y the plant output and d the disturbances.

If the model $\hat{P}$ is an approximate description of the real, but incompletely known plant P, one way to cope with uncertainty can be by using a multiplicative type of description such as

$$P(i\omega) = \hat{P}(i\omega)(1 + l_m(i\omega)) \qquad \text{with} \qquad |l_m(i\omega)| \le \bar{l}_m(\omega)$$

In order to satisfy robust stability, the sensitivity function S and its complementary function T have to satisfy:

$$|T(i\omega)\; l_m(i\omega)| < 1\; \forall\, \omega \tag{6.12}$$

These conditions can be expressed as:

$$\|\hat{P}Ql_m\|_\infty < 1 \tag{6.13}$$

The key idea of the method presented here is to implement both the model of a plant and the controller with interpolation models, such as the one described above (see Figure 6.8).

In the frequency-based IMC control structure, the error signal is fed to a bank of *comb* filters where it is separated into its spectral components. Each of the components is then multiplied by a complex gain and fed to the process and model.

If there are no modelling errors, the sensitivity function and its complementary are given by:

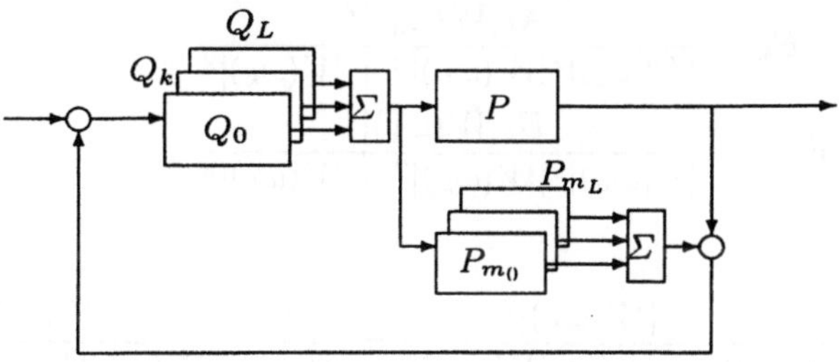

Fig. 6.8. IMC realization with frequency bands

$$\begin{aligned} S(\omega) &= 1 - P(\omega)Q(\omega) & (6.14) \\ T(\omega) &= P(\omega)Q(\omega) & (6.15) \end{aligned}$$

Both functions play an important role in assuring robustness of the closed loop controlled system in presence of uncertainties. *Perfect control* would be achieved by making $|S|$ as small as possible and $|T|$ as near as possible to 1. If robust control considerations are taken into account $|T(\omega)|$ is limited by $1/|\bar{l}_m(\omega)|$. For high frequencies $|\bar{l}_m(\omega)|$ always exceeds unity, usually complete uncertainty existing about the phase, and so $|S|$ cannot be driven to zero. It is necessary to find a trade–off solution between these opposite objectives. A way of doing this can be by the optimization of the function proposed by Kwakernaak [66]:

$$\|J(i\omega)\|_\infty = \|\ |V(i\omega)S(i\omega)|^2 + |W(i\omega)T(i\omega)|^2\ \|_\infty \qquad (6.16)$$

$V(i\omega)$ and $W(i\omega)$ being weighting functions.

The minimization of function J ($\forall\, \omega$) would guarantee the minimization of $\|J\|_\infty$ and $\|J\|_2$. The frequency responses of the plant and controller at frequency ω_k can be denominated $P(i\omega_k) = A_k + iB_k$ and $Q(i\omega_k) = E_k + iF_k$. If no restrictions are imposed on E_k and F_k, that is, on the complex frequency response of the controller, and taking into account that the filter banks produce almost orthogonal signals [83], function J can be optimized independently at every sampling point ω_k. This can be accomplished by differentiating and solving $\frac{\partial J(\omega_k)}{\partial E_k} = 0$ and $\frac{\partial J(\omega_k)}{\partial F_k} = 0$. The values of function J (6.16) at the interpolation points can be calculated by:

$$\begin{aligned} J(\omega_k) &= |V_k|^2[(A_kE_k - B_kF_k)^2 + (A_kF_k + B_kE_k)^2] \\ &+ |W_k|^2[(1 - A_kE_k + B_kF_k)^2 + (A_kF_k + B_kE_k)^2] \qquad (6.17) \end{aligned}$$

where $V_k = V(\omega_k)$ and $W_k = W(\omega_k)$. This leads, calculating the partial derivatives, to:

$$E_k = \frac{A_k |W(\omega_k)|^2}{|P(\omega_k)|^2 |W(\omega_k)|^2 + |V(\omega_k)|^2} \tag{6.18}$$

$$F_k = -\frac{B_k |W(\omega_k)|^2}{|P(\omega_k)|^2 |W(\omega_k)|^2 + |V(\omega_k)|^2} \tag{6.19}$$

That is:

$$Q(\omega_k) = \frac{P^*(\omega_k)}{|P(\omega_k)|^2} \frac{|W(\omega_k)|^2}{|W(\omega_k)|^2 + |V(\omega_k)|^2} = \frac{1}{P(\omega_k)} \frac{|W(\omega_k)|^2}{|W(\omega_k)|^2 + |V(\omega_k)|^2} \tag{6.20}$$

Notice that if there are no uncertainties in the model, perfect control would be achieved by making $Q(\omega) = 1/P(\omega)$. When robust control considerations are taken into account, the *perfect control* solution is weighted by the functions $V(\omega)$ and $W(\omega)$. The closed loop frequency response is given by:

$$G_{cl}(\omega_k) = \frac{|W(\omega_k)|^2}{|W(\omega_k)|^2 + |V(\omega_k)|^2} \tag{6.21}$$

Another design method consists of applying a determined frequency response of the controlled system at the chosen interpolating points, as is done in this chapter, to obtain an adaptive controller. The main idea of the method is to implement both the plant model (with antiresonance characteristics) and the controller with interpolating models. In order to implement the frequency-based adaptive controller, the frequency response of the plant is needed at all interpolating points. The frequency response at these points can be obtained as shown in [26] by using banks of band filters for the input and output of the plant. For the estimation of the parameters describing these simple models of the plant in each interpolating frequency, a common recursive least squares algorithm can be used (using a regressor in each band). The standard recursive least squares algorithm with a variable forgetting factor has been demonstrated to have good properties when the number of estimated parameters is small. If the dynamic information entering the estimator is poor supervisory levels must be included, as mentioned in chapter 4.

The usual way of implementing the algorithm is by obtaining the frequency response of the plant from the estimated parameters, inverting equations (6.8) and (6.9), and from there, calculating the controller frequency response, using equation (6.20). In the case presented, the desired frequency response of the compensated system at the interpolating points (G_d) is imposed, and so the controller frequency response is obtained from $Q(w_k) = G_d(w_k)/P(w_k)$. In both cases, having the frequency response at the interpolating points $Q(w_k)$, the controller coefficients are then obtained by applying equations (6.8) and (6.9). Figure 6.9 shows the general control scheme proposed. In order to implement the controller, a delay of one sampling period has to be taken into account and must be cascaded within the process model (this is always possible in systems deriving from sampling a causal continuous process where the system output is read before sending the control signal) to avoid direct transmission of signals from the controller to the model output.

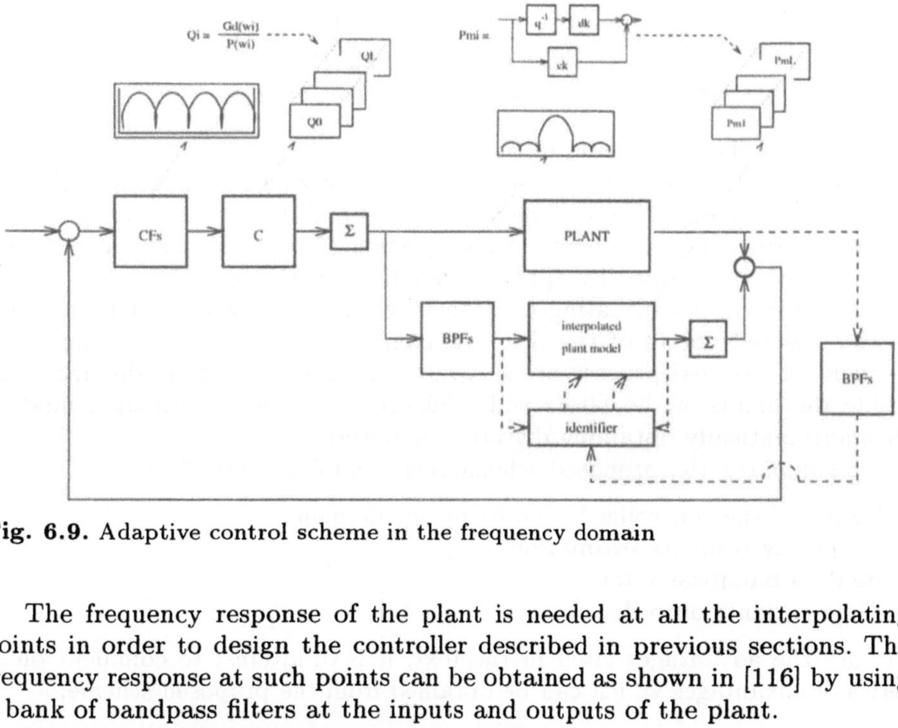

Fig. 6.9. Adaptive control scheme in the frequency domain

The frequency response of the plant is needed at all the interpolating points in order to design the controller described in previous sections. The frequency response at such points can be obtained as shown in [116] by using a bank of bandpass filters at the inputs and outputs of the plant.

If the interpolated transfer function of the plant is given by:

$$P_k(z) = \sum_{k=0}^{L} (a_k + b_k z^{-1}) H_k(z) \tag{6.22}$$

in order to estimate plant model parameters, the plant representation is parameterized in:

$$\begin{aligned} \boldsymbol{\theta}_k &\doteq [a_k \; b_k]^T \\ y_k(t) &= \boldsymbol{\theta}_k \boldsymbol{\phi}_k(t-1) + d_k(t) \end{aligned} \tag{6.23}$$

$\boldsymbol{\phi}_k$ being the estimator in band k defined as

$$\boldsymbol{\phi}_k(t) \doteq [u_k(t) \; u_{k-1}(t)]^T \tag{6.24}$$

The estimation of $\boldsymbol{\theta}_k(t)$ can be updated with the recursive least squares estimation algorithm:

$$\hat{\boldsymbol{\theta}}_k(t) = \hat{\boldsymbol{\theta}}_k(t-1) + \mathbf{L}_k(t)\epsilon_k(t) \qquad \text{for} \qquad k = 0, \cdots, L \tag{6.25}$$

where $\epsilon_k(t)$ is the estimation error in band k,

$$\epsilon_k(t) \doteq y_k(t) - \hat{\theta}_k^T(t-1)\phi_k(t-1) \tag{6.26}$$

The adaptation gain $\mathbf{L}_k(t)$ and covariance matrix $\mathbf{P}_k(t)$ are calculated as:

$$\mathbf{L}_k(t) = \frac{\mathbf{P}_k(t)\phi_k(t-1)}{c(t) + \phi_k^T(t-1)\mathbf{P}_k(t)\phi_k(t)} \tag{6.27}$$

$$\mathbf{P}_k(t+1) = [\mathbf{I} - \mathbf{L}_k(t)\phi_k(t)]\mathbf{P}_k(t)/c(t) \tag{6.28}$$

where $c(t)$ is the variable forgetting factor.

The frequency response of the plant can easily be obtained from the estimated parameters, inverting equations (6.8) and (6.9), and from there, the controller frequency response, using equation (6.20). The controller coefficients are then obtained by applying equations (6.8) and (6.9).

In this way, by estimating a parametric model of low order for each one of the selected points of the Nyquist diagram, an estimation scheme independent of the real process order is obtained. Moreover, only disturbances inside the bands of the filters will influence the different estimated modes, thus automatically obtaining disturbance rejection.

In summary, the proposed scheme includes (as done in [56]):

- Design of the controller in the frequency domain.
- Frequency response estimation.
- Bank of bandpass filters.
- Time-variant controller.

Besides the advantages cited in the text, it is of interest to comment on a series of advantages which can be obtained from the proposed scheme:

- The estimation is concentrated in frequencies used for design purposes. Disturbances with frequency components out of these bands are filtered.
- It is not necessary to over excite in cases with processes of high order because only a few points of the Nyquist curve are estimated.
- Low order models can be used to control high order processes.
- The order of the process need not be known.
- The control scheme can respond to changes in plant dynamics.

Stability analysis

There are two sources of modelling errors in the control scheme presented:

1. The incomplete knowledge of the process inherent to all adaptive control schema.
2. The approximation errors due to the frequency interpolation mechanism used.

The model $\hat{P}$ is affected by both error sources while the controller Q is only affected by the second source.

These errors (or uncertainties) can be described by the following equations:

$$P = \hat{P}(\omega)(1 + l_m(\omega)) \tag{6.29}$$

$$Q = \frac{G_d(\omega)(1 + l_c(\omega))}{\hat{P}(\omega)} \tag{6.30}$$

The closed loop characteristic polynomial is then given by

$$1 + G_d(\omega)l_m(\omega)(1 + l_c(\omega))$$

The controller approximation error, $l_c(\omega)$, can be kept small by choosing N to be sufficiently large. The uncertainty or modelling error $l_m(\omega)$ can be made small for those bands with persistent excitation. Notice that at those points where l_m cannot be kept small, because of a lack of excitation in the corresponding band, robustness will be guarantee by making $G_d(\omega)$ small enough to satisfy inequality (6.13).

Regarding these comments, two main drawbacks have been encountered in the proposed control scheme:

- On one hand, the existence of antiresonance characteristics causes the identification of the frequency response of the plant near these modes to reach very low values. Due to the fact that the control scheme obtains the control signal using the inverse of the estimated plant, very high values can be obtained of the control signal. A supervising scheme is shown in the sections devoted to commenting the results obtained with this control scheme.
- On the other hand, the choice of the transfer function which provides the desired behaviour of the controlled system and, particularly, the choice of the desired magnitude and phase values of the system near the antiresonance modes is critical, leading to different kind of behaviour in function of the chosen values.

6.1.6 A case study: linear system with one antiresonance mode

In order to show the advantages of using the method described above, some studies were made using a simple linear model with one modelled antiresonance mode. The transfer function of the model corresponds (in the continuous time domain) to:

$$G(s) = \frac{s^2 + 2\delta w_n s + w_n^2}{(s + a)^3}$$

where $w_n = 1, \delta = 0.01$ and $a = 1$. The discrete transfer function, taking into account the zero order hold and with a sampling time of 0.25 seconds, is given by:

$$G(z) = \frac{.173z^{-1} - .3339z^{-2} + .1717z^{-3}}{1 - 2.3364z^{-1} + 1.8196z^{-2} - .4724z^{-3}}.$$

The interpolation using Lagrange filters was made by using N=23 (L=11) interpolation points and choosing only the first three (L_1 = 3) for control purposes. In this way, information about the magnitude and phase of the interpolated system at low frequencies, near the antiresonance mode, and at medium-high frequencies, is obtained, and only three simple identifications have to be performed at each sampling time.

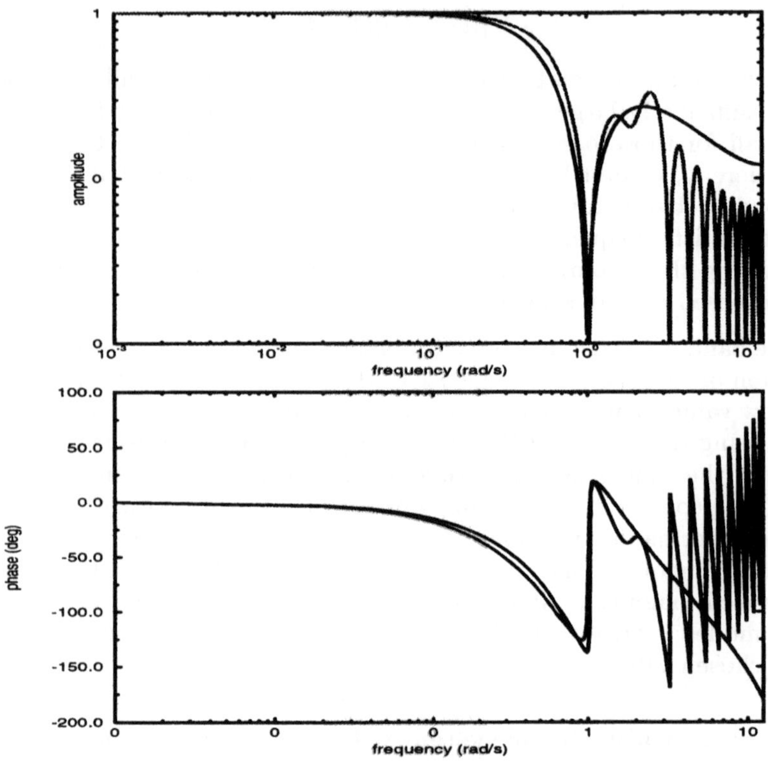

linear system: solid-line; interpolated system: dotted-line

Fig. 6.10. Frequency response of the antiresonance linear system and the interpolated one

The frequency response of the system and that obtained by using the Lagrange filters can be seen in Figure 6.10. The unitary step response in open loop can be seen in Figure 6.11, where the presence of the antiresonance mode slightly appears in the response.

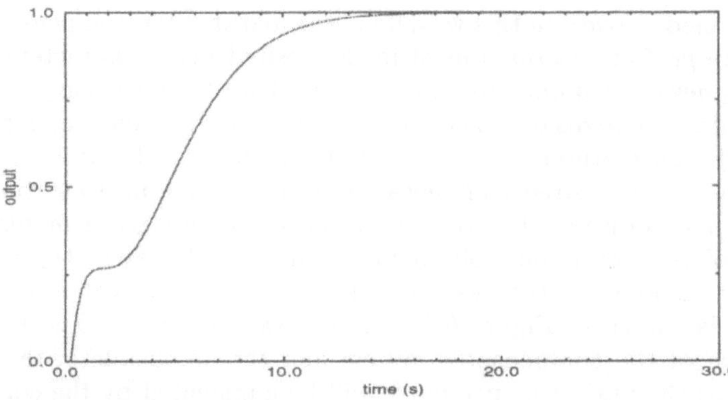

Fig. 6.11. Step response

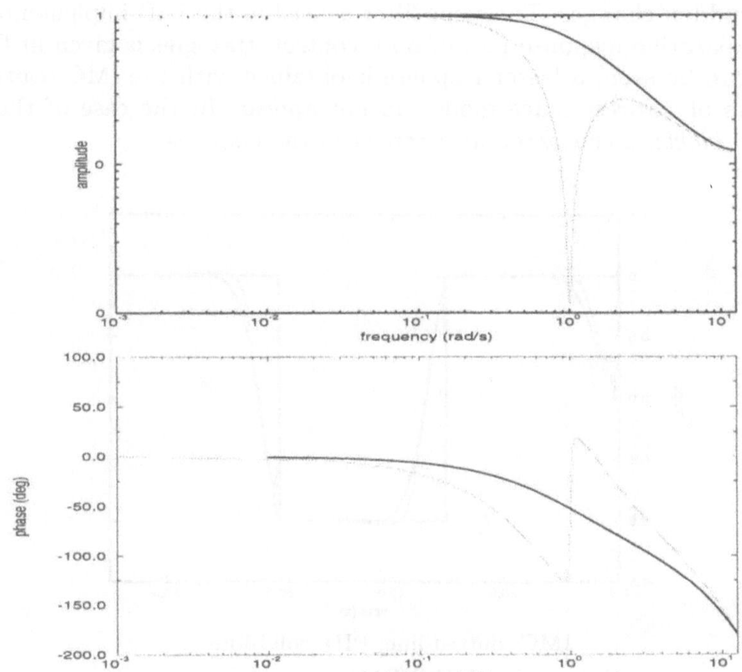

dotted line: system; solid line: frequency domain specifications

Fig. 6.12. Frequency response of the system and specifications

The implementation of the frequency-based IMC could be performed by using the weighted inverse of the linear plant at each interpolating frequency (as mentioned above). In this way, it is expected that the compensated system will assure *perfect control*, almost in the desired phase characteristics. From the robustness viewpoint, this can be a drawback because heavy demands are made on the compensated system response. As has been mentioned before, another design method can be by simply imposing a determined amplitude and phase of the desired compensated system on the interpolating frequencies. In this example, the control specifications are given in terms of the location of the dominant pole of the compensated system ($s_d = -1$). The frequency response of the system and that of the one that meets the specifications is shown in Figure 6.12. In this way, it can be expected that the bandwidth of the compensated system and the amplitude of the frequency response at the antiresonance mode will be augmented by the compensator.

In order to perform a comparative study of the behaviour of the proposed algorithm, an analytical PID showed in the book of Shahian and Hassul [111] was designed to control the prescribed system (taking the mentioned specification into account). As the controller gain at the antiresonance frequency can achieve large values, the output of the bank controllers is filtered in order to avoid sudden changes. The same filter is used in the PID implementation.

A comparative response plot of both control strategies is given in Figure 6.13. As can be seen, a faster response is obtained with the IMC controller, and effects of antiresonance modes do not appear. In the case of the PID controller, *kickback* characteristics appear in the response.

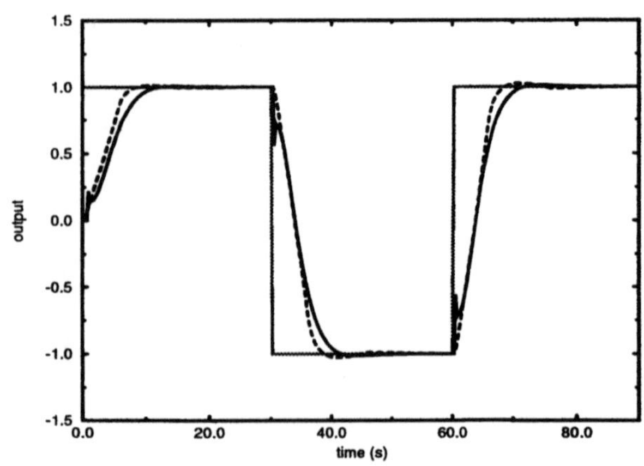

IMC: dotted-line; PID: solid-line

Fig. 6.13. Comparative results IMC-PID

Robustness of the proposed scheme

If both the controlled system sensitivity function and the controller properties are supervised, it is possible to obtain a design methodology to modify the behaviour of the controlled system increasing its robustness.

One of the main problems presented by the proposed scheme is that it can lead to unacceptable results in those cases in which the identification carried out at the chosen frequencies is not accurate, mainly near the antiresonance mode, because of the great value of the control signal which can be generated: $Q(\omega_k) = G_d(\omega_k)/\hat{P}(\omega_k)$ ($Q(\omega_k)$ being the frequency response of the controller, $G_d(\omega_k)$ the expected frequency response of the plant and $\hat{P}(\omega_k)$ the estimated plant in each point). In order to compensate for these possible problems, an external loop is included in the control scheme as follows.

If the linear system treated in this section in its adaptive version is implemented, after an initial stage in which the parameters are adjusted by introducing sine wave signals of different frequencies, the response represented in Figure 6.14 can be obtained. The evolution of the control signals in each band and that of the a_i parameters of each one of the simplified identification models ($P_k(z) = \sum_{k=0}^{L}(a_k + b_k z^{-1})H_k(z)$) can be analyzed in Figure 6.15.

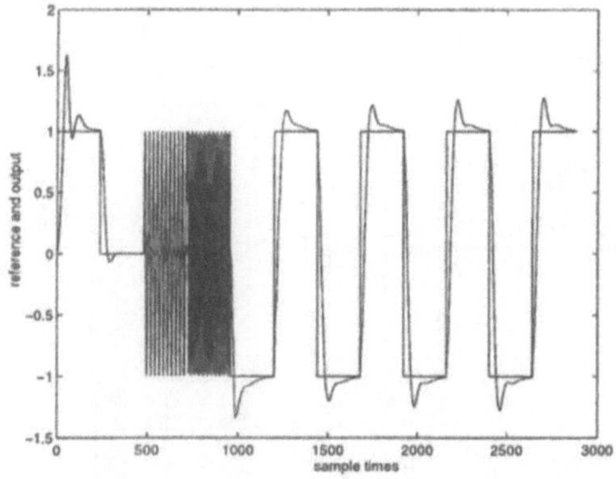

Fig. 6.14. Simulation with the adaptive frequency-based IMC controller

The proposed mixed scheme of supervision and control is shown in Figure 6.16. It mainly consists of a reference model in parallel which adjusts to the desired response if deviations of the temporal response from the expected trajectory are observed. That is, a comparison between the output of the model used to impose a determined frequency response at selected points and the real response is performed; if deviations are detected, the desired frequency response is modified according to a certain adaptation law.

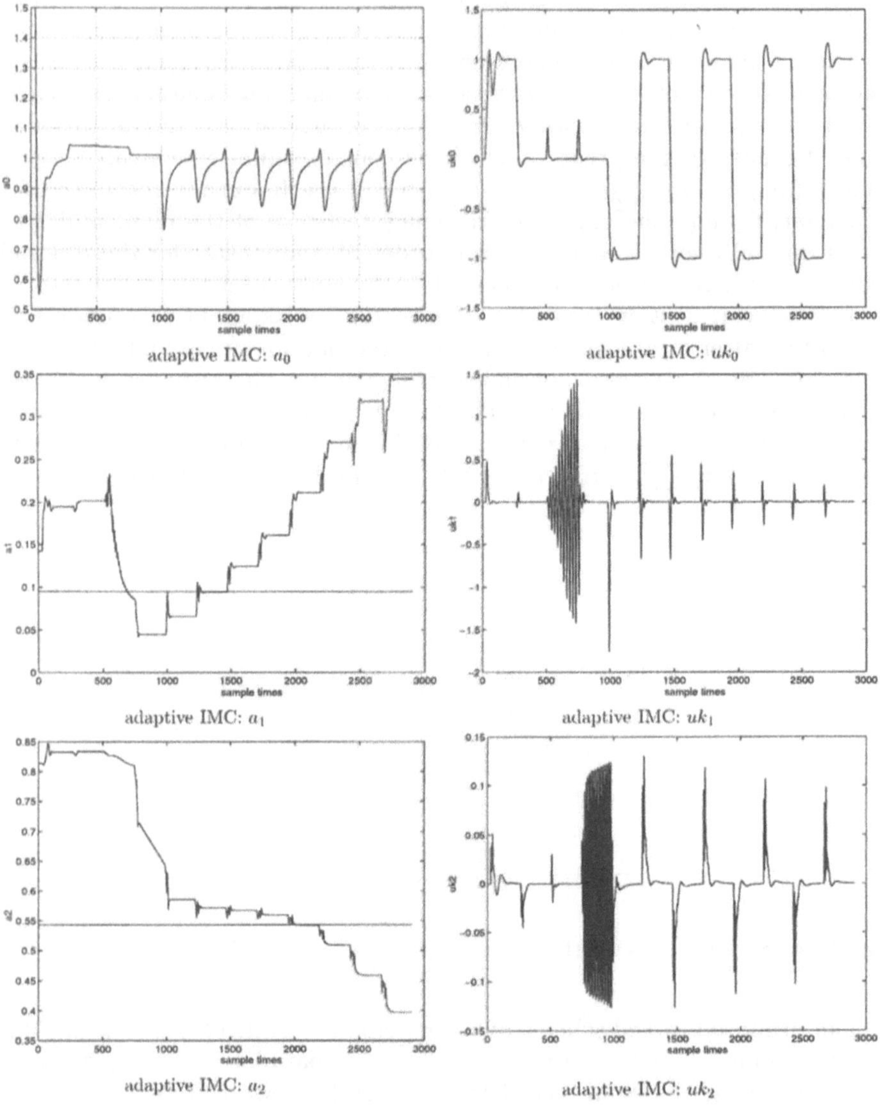

adaptive IMC: a_0 adaptive IMC: uk_0

adaptive IMC: a_1 adaptive IMC: uk_1

adaptive IMC: a_2 adaptive IMC: uk_2

Fig. 6.15. Simulation with the adaptive frequency-based IMC controller

In the mentioned case, a simple constant ($Q(\omega_k) = \hat{G}_d(\omega_k)/\hat{P}(\omega_k)$, with $\hat{G}_d(\omega_k) = \gamma G_d(\omega_k)$) has been used, but this scheme can utilize any of the classical design Model Reference Adaptive Control (MRAC) criteria, such as the MIT rule.

The improvements obtained in respect to the normal scheme with the introduction of the supervisory loop can be seen in Figure 6.17, in this way allowing for deviations to be diminished in respect to the nominal behaviour.

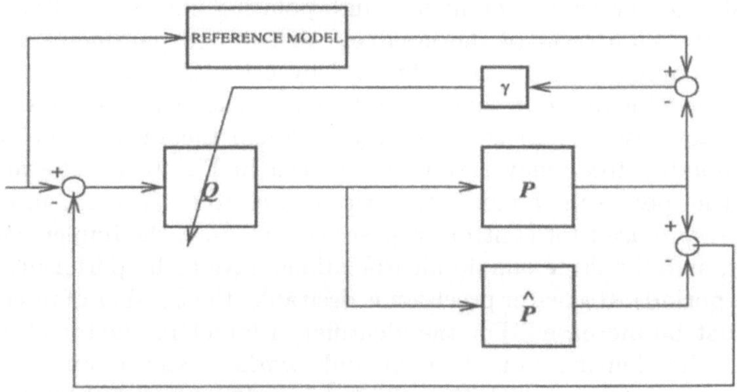

Fig. 6.16. Mixed control scheme

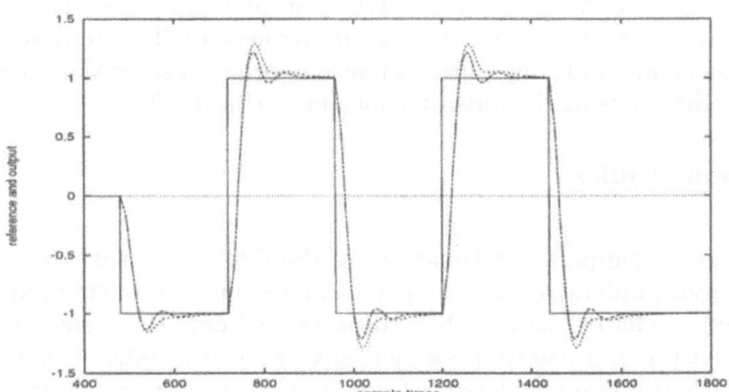

dashed line: response with mixed scheme; dotted line: normal response

Fig. 6.17. Response using the mixed scheme

6.1.7 Application to the distributed solar collector field

This section is concerned with the application of the frequency-based IMC to the control of the ACUREX distributed solar collector field. The controller has been implemented using the feedforward controller in series with the plant in order to diminish the influence of disturbances on the estimated frequency response of the plant.

The difficulty inherent to the implementation of an adaptive control strategy when coping with high order systems has been mentioned in other chapters. Due to this fact, the approximation described here can be used to take into account system dynamics in a wide frequency range, decomposing the problem into identifying a small set of low order models.

In order to choose the number of interpolating points, the frequency response of the plant and of the nonlinear distributed parameter model was used to be compared with that obtained by using the Lagrange filters. For example, the plot of the frequency response of one of the models (in which the frequency response at $\omega = 0.02$ rad/s should theoretically be zero) and the interpolated frequency response is shown in Fig. 6.18. The number of interpolating points in the unit circle was chosen to be $N = 11$, and only the first three were used for control purposes (in this way, the implementation is very easy, as only three simple identifications have to be performed at each sampling period). If greater precision is desirable, the number of interpolating points must be increased. For the identifier, a forgetting factor of $c = 0.975$ was used. The identification algorithm only works when the input signal contains dynamic information (otherwise the forgetting factor is made equal to one and the covariance matrix is reset, making its non-diagonal values equal to zero).

As in the previous section with the linear model, the desired behaviour of the controlled system was chosen in terms of the magnitude and phase of the compensated system at the interpolating frequencies. The desired behaviour was chosen so as to obtain a fast response in order to show that even in such cases, the influence of the antiresonance modes is small.

Simulation studies

In order to compare the behaviour of the IMC controller with other control strategies implemented at the plant, some simulations using the nonlinear distributed model of the field have been carried out. Here, some results are presented using the atmospheric and environmental data of one day's operation with the plant (reference test). The chosen design parameters have been:

- Number of interpolating points $N = 11$.
- Number of points at the positive semi-circumference $L = (N-1)/2 = 5$.
- Number of points used for controller design purposes: $L_1 = 3$.
- Interpolating frequencies: $\omega = (2\pi k)/(NT)$ (en rad/s), $T = 39$ seconds being the sampling time.
- Chosen design frequencies: $\omega_{0,1,2} = \{0, 0.0146, 0.029\}$ (rad/s).

- Specifications in the design frequencies:
  - Magnitude: $\{1, 0.3785, 0.19\}$.
  - Phase: $\{0, -1.47, -1.81\}$ (rad.).

These specifications have been chosen analyzing the frequency response of a linear system with the desired behaviour at those points.

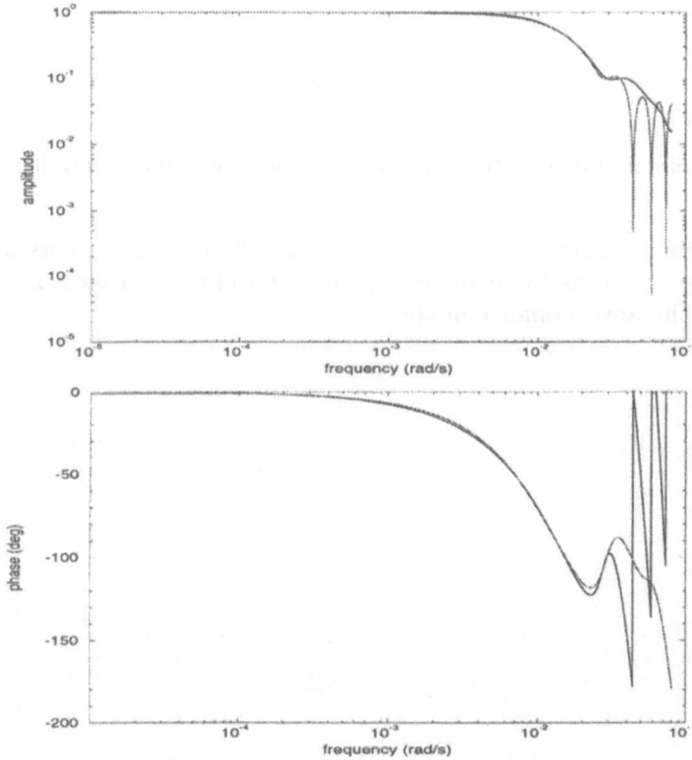

Fig. 6.18. Frequency response of the plant model and the interpolated frequency response

Figure 6.19 shows the result obtained in a simulation with data of the reference test. As can be seen, good results are obtained in the simulation without effects of the antiresonance modes appearing in the output. The evolution of one of the estimated parameters a_1 (corresponding to the second interpolating frequency) is shown as an example in the same figure.

Figure 6.20 shows the Nyquist plot corresponding to the four models of the frequency response (obtained for different operating conditions, as analyzed in chapter 5). The temporal evolution of the three estimated points of the Nyquist curve are superimposed. As can be seen, these points evolve from one operating point (medium-high flow) to another operating point (medium-low

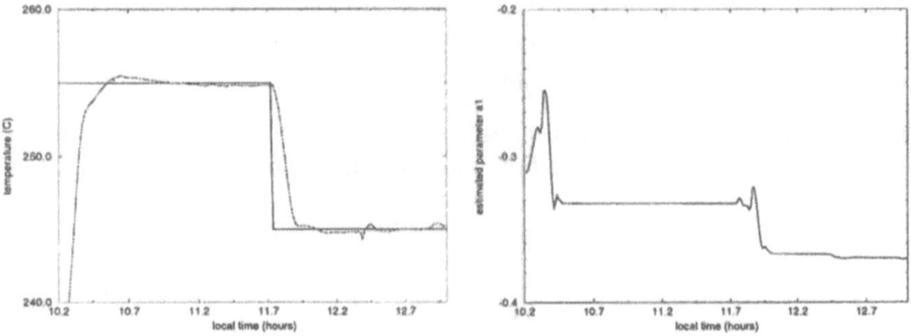

set point and outlet oil temperature (°C) evolution of parameter a_1

Fig. 6.19. Simulation with the adaptive frequency-based IMC controller

flow) covering a range from 7 l/s to 4 l/s (oil flow). There exists a greater discrepancy at the second estimated point due to the fact that it is the point nearest to the antiresonance mode.

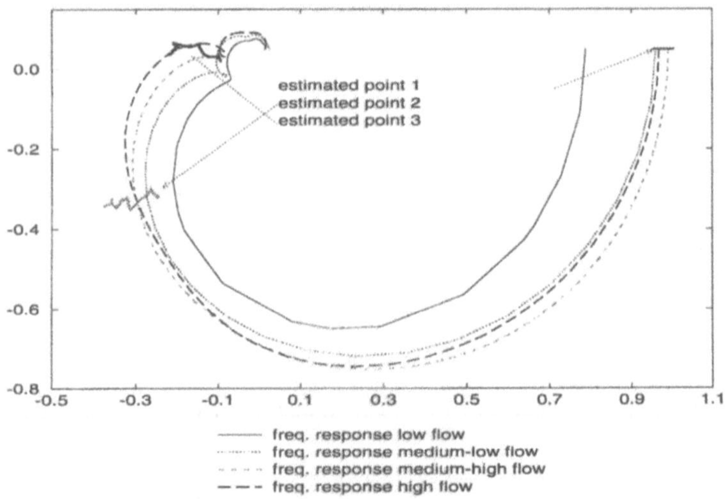

Fig. 6.20. Temporal evolution of the points in the Nyquist plot

Finally, in Figure 6.21, a comparison of the result obtained when using the algorithm with FIR models for identification and that obtained when using the first order model can be seen.

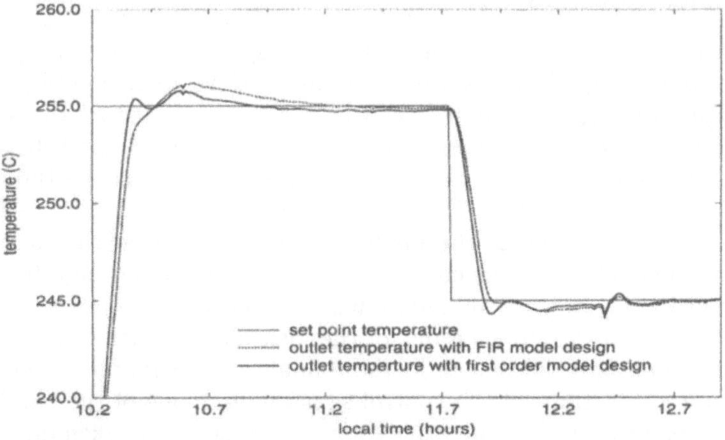

Fig. 6.21. Response with different estimation models

Plant results

Two tests have been performed with the proposed control scheme. The chosen design values have been the same as those used in the simulation studies. In Figures 6.22 and 6.23 the results obtained during an operation with strong disturbances in the solar radiation levels are presented. Figure 6.22 shows the evolution of the characteristic variables of the test (set point, inlet and outlet oil temperatures, direct solar radiation and oil flow), while in Figure 6.23 the evolution of inputs uk_i in each band and that corresponding to parameters a_i of the estimation models is presented.

Figures 6.24 and 6.25 show another test carried out with the distributed solar collector field. It corresponds to a day with high radiation, but with occasional passing clouds which produce significant changes in the solar radiation levels. The tracking characteristics are acceptable, presenting a response time greater than those obtained with other control schema tested at the plant (due to a non-optimal selection of the specifications). The outlet oil temperature shows oscillations due, on one hand, to solar radiation changes and on the other to a possibly wrong estimation of the frequency corresponding to the first antiresonance mode or to the necessity of taking into account more points of the Nyquist curve in the design of the controller to better estimate the position of the first mode. Figure 6.25 presents the evolution of the characteristic parameters of the simple identification models in the selected frequency bands and the control signals corresponding to each band.

Disturbance rejection

The strong disturbances which affect the system in the tests presented cause the control signal to be saturated in parts of the operation (Figure 6.22) and to experiment great changes due to solar radiation disturbances (Figure 6.24). With this control system, it is to be expected that good disturbance rejection capabilities should be shown by the system, due to the filtering action of the bandpass filters. Nevertheless, as can be seen in Figure 6.23, the amplitude and frequency of the disturbances is so diverse, that substantial changes in the control signals corresponding to medium and high frequencies are produced (uk_1 and uk_2 in Figure 6.23). After analyzing these results, it is useful to incorporate to each one of the identifiers in each frequency band a number of heuristic rules such as those analyzed in chapter 4, and to take into account saturations of the control signal in each one of the bands of the controller (uk_i) to avoid the identification of dynamics not taken into account by the simplified models.

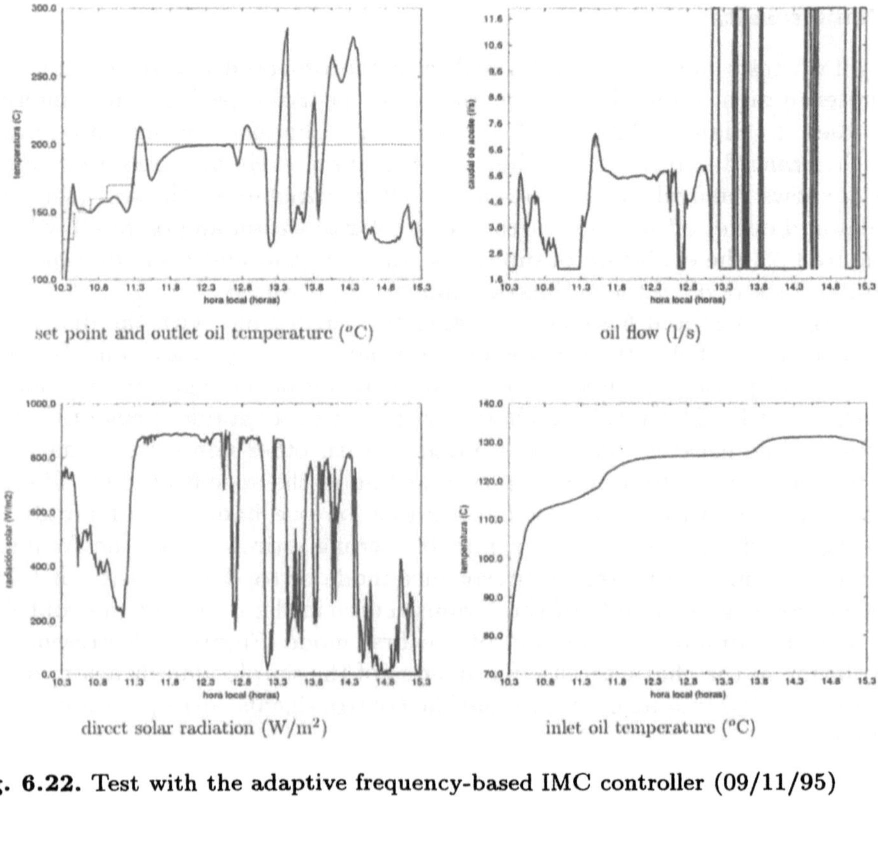

Fig. 6.22. Test with the adaptive frequency-based IMC controller (09/11/95)

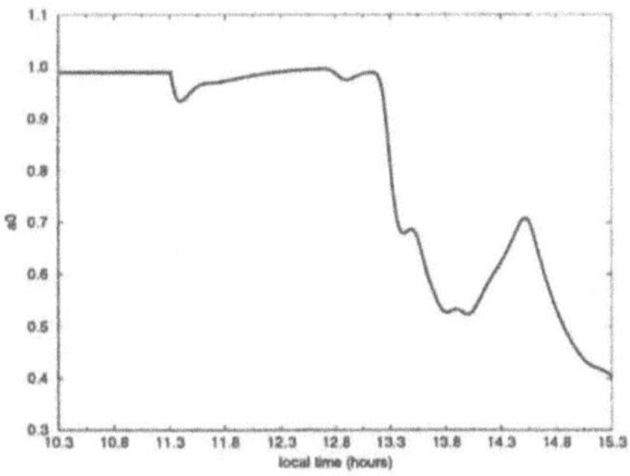

parameter a_0

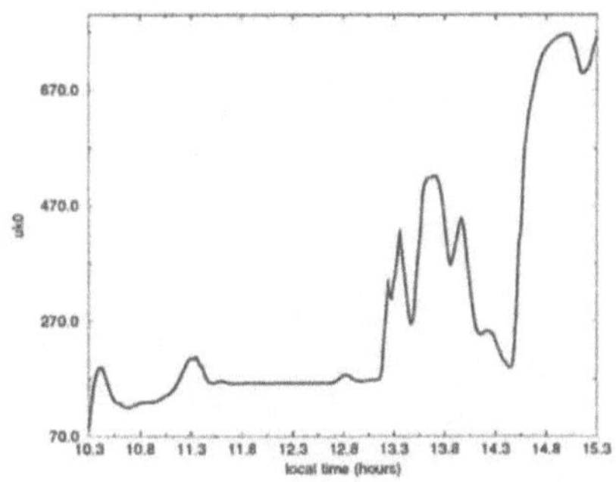

evolution of uk_0

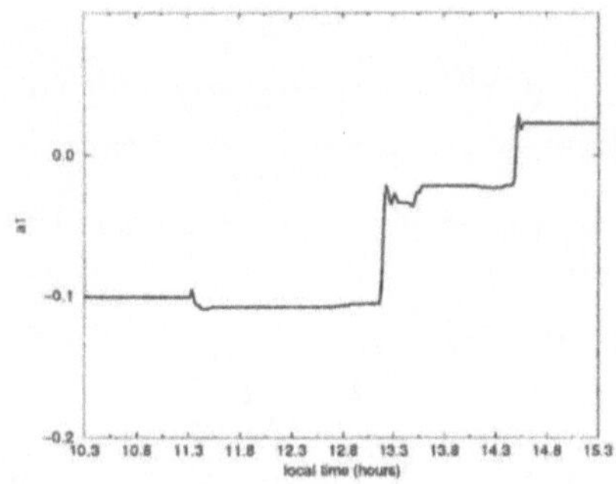

parameter a_1

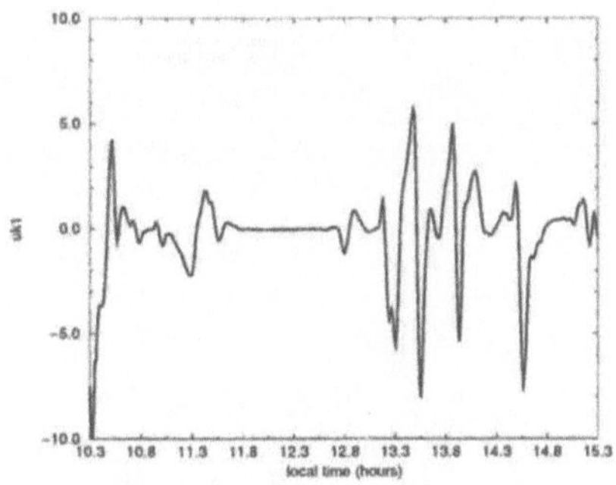

evolution of uk_1

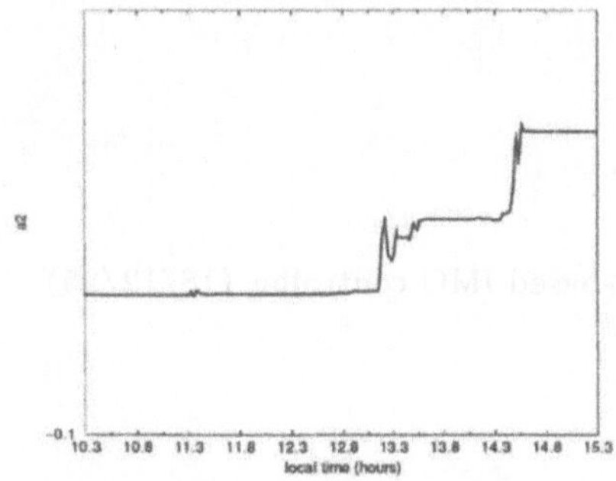

parameter a_2

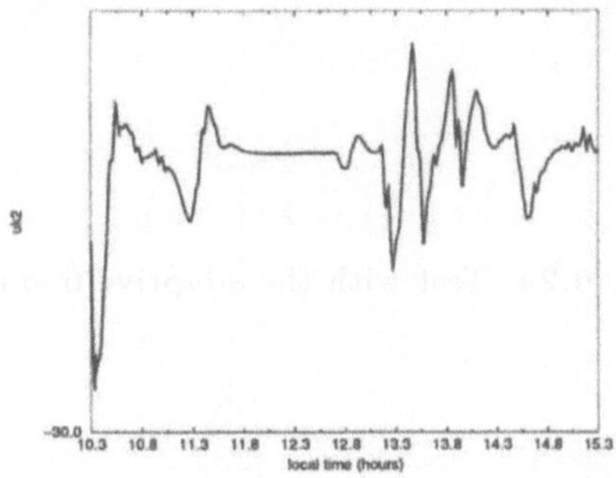

evolution of uk_2

Fig. 6.23. Test with the adaptive frequency-based IMC controller (09/11/95)

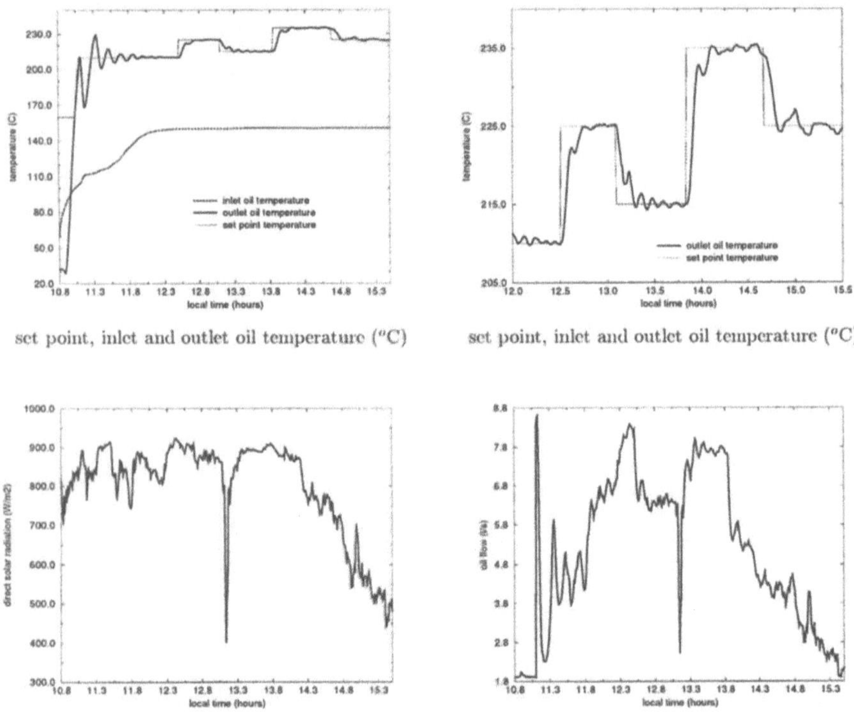

set point, inlet and outlet oil temperature (°C)   set point, inlet and outlet oil temperature (°C)

direct solar radiation (W/m^2)   oil flow (l/s)

Fig. 6.24. Test with the adaptive frequency-based IMC controller (18/12/95)

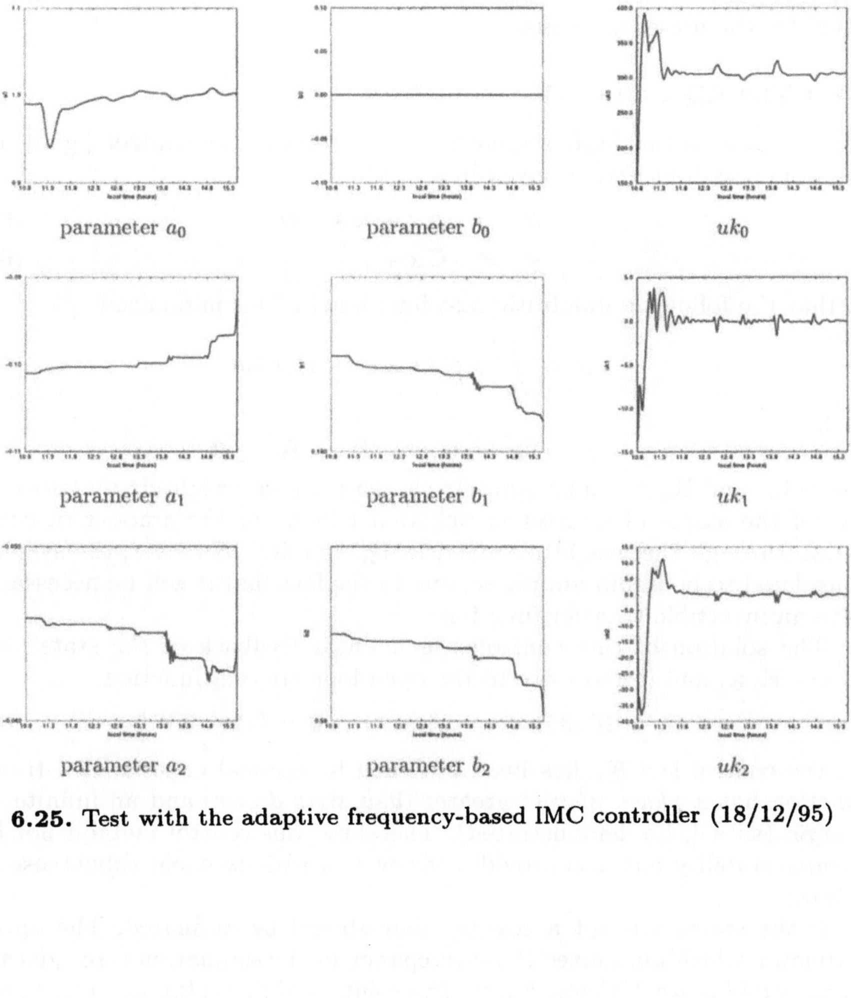

Fig. 6.25. Test with the adaptive frequency-based IMC controller (18/12/95)

6.2 Linear Quadratic Gaussian Optimal Control (LQG)

6.2.1 Introduction

The problem of finding a control law for a linear system, with measurable noise acting on it, can be solved by minimizing a criterion which measures how well the closed loop behaves. In particular, by using a weighted quadratic function of the states and the control law, the resulting controller is a linear optimal one. This is a well–known *Linear Quadratic Regulator* (LQR) if noises are not taken into account and a *Linear Quadratic Gaussian Controller* (LQG) if the noises are gaussian.

6.2.2 The LQR and LQG regulators

The purpose of the LQR regulator is to determine the control signal $\mathbf{u}$ for the time invariant system given by:

$$\dot{\mathbf{x}} = \mathbf{A}\mathbf{x} + \mathbf{B}\mathbf{u} + \mathbf{w} \tag{6.31}$$

$$\mathbf{y} = \mathbf{C}\mathbf{x} + \mathbf{v} \tag{6.32}$$

so that the following quadratic cost functional ($\mathcal{J}$) is minimized:

$$\mathcal{J} = \int_0^\infty (\mathbf{x}^T \mathbf{Q}_c \mathbf{x} + \mathbf{u}^T \mathbf{R}_c \mathbf{u})\, dt \tag{6.33}$$

with:

$$\mathbf{Q}_c = \mathbf{Q}_c^T \geq \mathbf{0}\,, \qquad \mathbf{R}_c = \mathbf{R}_c^T \geq \mathbf{0}$$

where $\mathbf{Q}_c$ and $\mathbf{R}_c$ are appropriately chosen constant weighting matrices. The size of the states of interest is weighted relative to the amount of control action through the weighting matrices $\mathbf{Q}_c$ and $\mathbf{R}_c$. Moreover, the system is considered to be minimum phase, due to the fact that it will be necessary to have an invertible transfer function.

The solution of this controller is a linear feedback of the state vector ($\mathbf{u} = -\mathbf{K}_c\mathbf{x}$) and it gives rise to the open loop transfer function:

$$G_c = \mathbf{K}_c \boldsymbol{\Phi} \mathbf{B} \qquad \text{with} \qquad \boldsymbol{\Phi} = (s\mathbf{I} - \mathbf{A})^{-1} \tag{6.34}$$

As the control law $\mathbf{K}_c$ has been obtained by optimal criteria, this transfer function has a *phase margin* greater than *sixty degrees* and an infinite *gain margin* (as will be demonstrated). Therefore, this control method not only ensures stability but also provides the system with excellent robustness conditions.

If the states are not accessible they should be estimated. The optimal estimator which minimizes the discrepancy in the estimation error gives rise to the well known *Kalman Filter*. The solution of this estimator proves to be a dual problem to that of the control problem (LQR).

The LQG consists of solving the problem of optimal control and estimation. The solution to this problem is based on the Separation Theorem, according to which the control strategy can be divided into two parts:

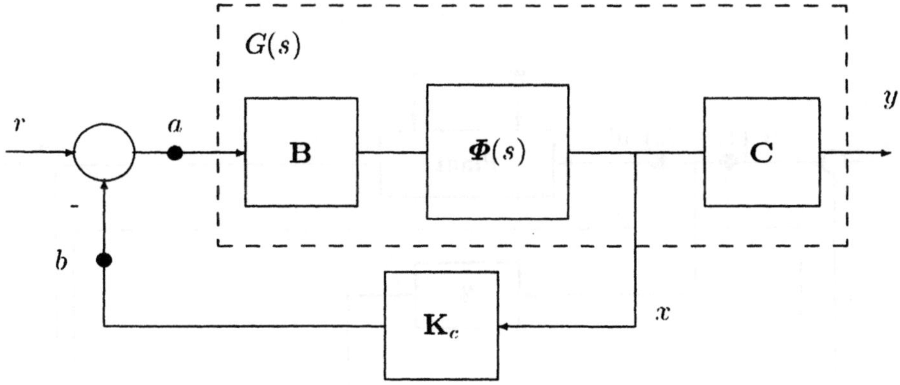

Fig. 6.26. Feedback of the state vector

1. An optimal control problem, from which the regulation of state variables by feedback is obtained:

$$\begin{aligned} \mathbf{u} &= -\mathbf{K}_c\hat{\mathbf{x}} \\ \text{with} \quad \mathbf{K}_c &= \mathbf{R}_c^{-1}\mathbf{B}^T\mathbf{P}_c \end{aligned}$$

   where $\mathbf{P}_c$ is the unique, symmetric, positive semidefinite solution of the algebraic *Riccati equation*:

$$\mathbf{A}^T\mathbf{P}_c + \mathbf{P}_c\mathbf{A} - \mathbf{P}_c\mathbf{B}\mathbf{R}_c^{-1}\mathbf{B}^T\mathbf{P}_c + \mathbf{Q}_c = \mathbf{0} \tag{6.35}$$

2. An optimal filtering problem, which solves the observation using the *Kalman filter*, whose corresponding dynamics are given by the system:

$$\frac{d\hat{\mathbf{x}}}{dt} = \mathbf{A}\hat{\mathbf{x}} + \mathbf{B}\mathbf{u} + \mathbf{K}_o(\mathbf{y} - \mathbf{C}\hat{\mathbf{x}}) \tag{6.36}$$

   where the optimal choice of $\mathbf{K}_o$, which minimizes $E[\mathbf{x} - \hat{\mathbf{x}}]^T[\mathbf{x} - \hat{\mathbf{x}}]$ is given by:

$$\mathbf{K}_o = \mathbf{P}_o\mathbf{C}^T\mathbf{R}_o^{-1} \tag{6.37}$$

   and $\mathbf{P}_o$ is the unique positive-semidefinite solution of the algebraic *Riccati equation*,

$$\mathbf{A}\mathbf{P}_o + \mathbf{P}_o\mathbf{A}^T + \mathbf{Q}_o - \mathbf{P}_o\mathbf{C}^T\mathbf{R}_o^{-1}\mathbf{C}\mathbf{P}_o = \mathbf{0} \tag{6.38}$$

   with

$$\begin{aligned} E\{\mathbf{w}(t)\mathbf{w}(\tau)^T\} &= \mathbf{Q}_o\delta(t-\tau) \\ E\{\mathbf{v}(t)\mathbf{v}(\tau)^T\} &= \mathbf{R}_o\delta(t-\tau) \end{aligned}$$

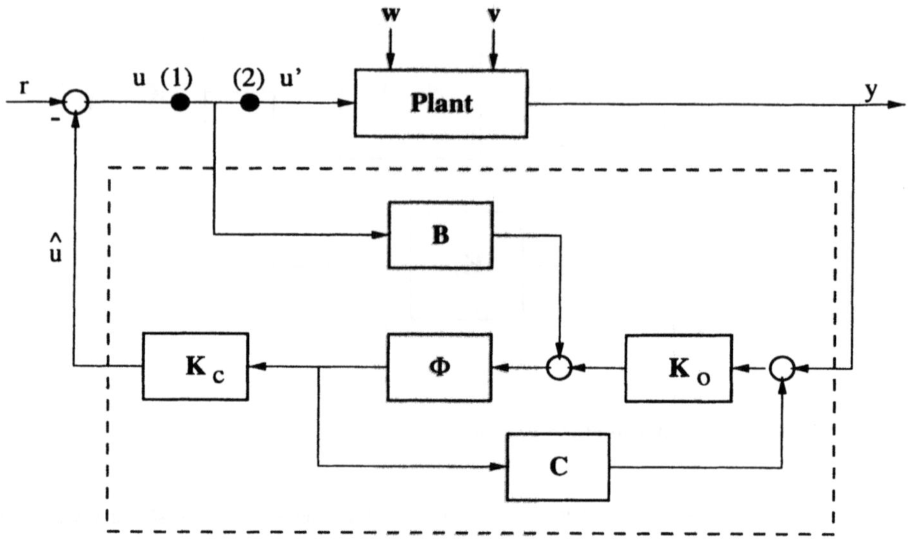

Fig. 6.27. Diagram of the LQG controller

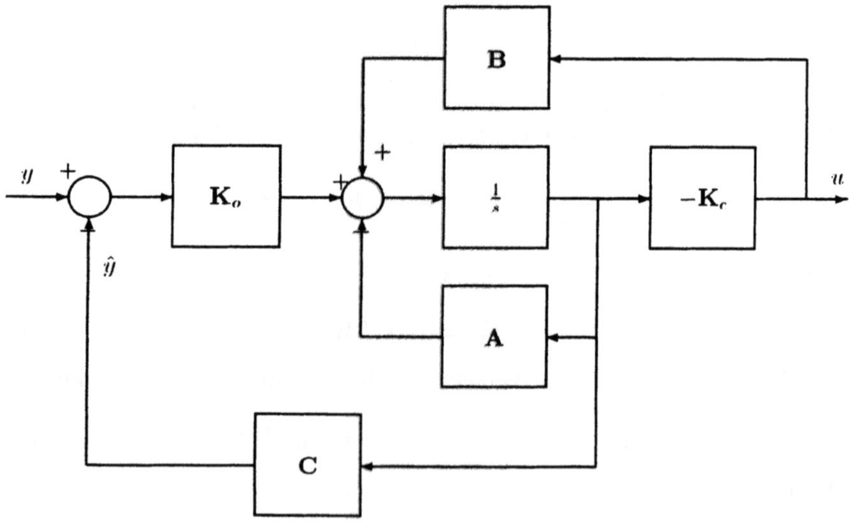

Fig. 6.28. LQG controller

The resulting regulation scheme is shown in Figure 6.27 and the resulting compensator is presented in Figure 6.28.

By mathematical manipulation, the resulting regulator can be obtained in terms of the transfer function. In this way, one obtains the classical control problem shown in Figure 6.29, valid for the regulation problem ($Ref = 0$).

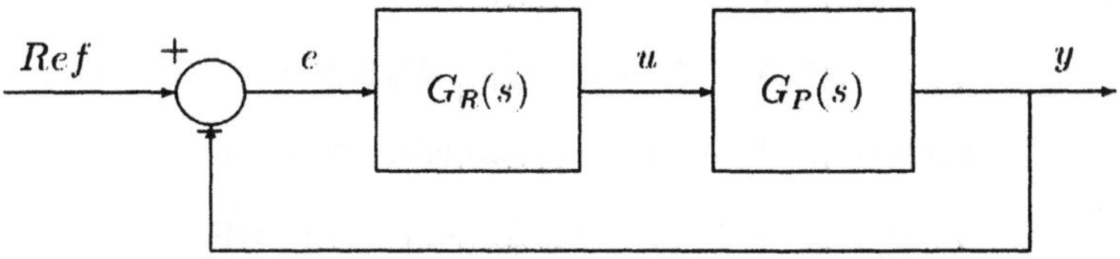

Fig. 6.29. Classical control problem

The transfer function of the regulator, and thus of the open loop $G_{LQG}(s)$, depends on the parameters of the *control law* and of the *Kalman filter*, which in turn depends on the *costs* assigned to the optimal control problem and the noise covariance matrices for the state estimation problem. It is a well known fact that the closed loop transfer function from the output y to the reference r when the state is accessible (LQR) is the same than that obtained when the Kalman filter is used to obtain the estimated states (LQG), so that the closed loop temporary specifications solved by the application of the control law will continue to be fulfilled. However, the open loop transfer function becomes:

$$G_{LQG}(s) = G_R(s)G_p(s) = \mathbf{K}_c(s\mathbf{I} - \mathbf{A} + \mathbf{BK}_c + \mathbf{K}_o\mathbf{C})^{-1}\mathbf{K}_o \ \mathbf{C\Phi B} \quad (6.39)$$

There is no fixed rule to guarantee the values which the new stability margins will have, but in most cases this variation usually goes into action by greatly deteriorating the phase margin of the compensated system, causing a consequent loss of robustness.

6.2.3 Establishment of the LQG method in the frequency domain

As shown by Doyle and Stein [42], [43], the *Riccati equation* can be interpreted in the frequency domain, so that expressions in terms of transfer functions can be obtained. Consider the system,

$$\dot{\mathbf{x}} = \mathbf{Ax} + \mathbf{Bu} \quad (6.40)$$

where $\mathbf{u}$ is considered to have dimension 1 for simplicity ($\mathbf{u} = u$), and under the performance criterion:

$$\mathcal{J} = \int_0^\infty (\mathbf{x}^T \mathbf{Q}_c \mathbf{x} + u^2)\, dt \tag{6.41}$$

Let $\mathbf{R}_c = 1$ without loss of generality, because it can be embedded in coefficients of $\mathbf{Q}_c$. The corresponding *Riccati equation* (equation 6.35), is given by:

$$\mathbf{A}^T\mathbf{P}_c + \mathbf{P}_c\mathbf{A} - \mathbf{P}_c\mathbf{B}\mathbf{R}_c^{-1}\mathbf{B}^T\mathbf{P}_c + \mathbf{Q}_c = \mathbf{0}$$

which can be rewritten as:

$$-\mathbf{P}_c\mathbf{A} - \mathbf{A}^T\mathbf{P}_c = \mathbf{Q}_c - \mathbf{P}_c\mathbf{B}\mathbf{B}^T\mathbf{P}_c$$

Adding and subtracting $s\mathbf{P}_c$ to the first member we have:

$$\mathbf{P}_c(s\mathbf{I} - \mathbf{A}) + (-s\mathbf{I} - \mathbf{A}^T)\mathbf{P}_c = \mathbf{Q}_c - \mathbf{P}_c\mathbf{B}\mathbf{B}^T\mathbf{P}_c$$

calling $\boldsymbol{\Phi}(s) = (s\mathbf{I} - \mathbf{A})^{-1}$ leads to:

$$\mathbf{P}_c\boldsymbol{\Phi}^{-1}(s) + \boldsymbol{\Phi}^{T^{-1}}(-s)\mathbf{P}_c = \mathbf{Q}_c - \mathbf{K}_c^T\mathbf{K}_c$$

Pre-multiplying by $\mathbf{B}^T\boldsymbol{\Phi}^T(-s)$ and post-multiplying by $\boldsymbol{\Phi}(s)\mathbf{B}$ we get:

$$\mathbf{B}^T\boldsymbol{\Phi}^T(-s)\mathbf{P}_c\underbrace{\boldsymbol{\Phi}^{-1}(s)\boldsymbol{\Phi}(s)}_{I}\mathbf{B} + \mathbf{B}^T\underbrace{\boldsymbol{\Phi}^T(-s)\boldsymbol{\Phi}^{T^{-1}}(-s)}_{I}\mathbf{P}_c\boldsymbol{\Phi}(s)\mathbf{B} =$$

$$\mathbf{B}^T\boldsymbol{\Phi}^T(-s)[\mathbf{Q}_c - \mathbf{K}_c^T\mathbf{K}_c]\boldsymbol{\Phi}(s)\mathbf{B}$$

$$\mathbf{B}^T\boldsymbol{\Phi}^T(-s)\underbrace{\mathbf{P}_c\mathbf{B}}_{\mathbf{K}_c^T} + \underbrace{\mathbf{B}^T\mathbf{P}_c}_{\mathbf{K}_c}\boldsymbol{\Phi}(s)\mathbf{B} =$$

$$\mathbf{B}^T\boldsymbol{\Phi}^T(-s)\mathbf{Q}_c\boldsymbol{\Phi}(s)\mathbf{B} - \mathbf{B}^T\boldsymbol{\Phi}^T(-s)\mathbf{K}_c^T\mathbf{K}_c\boldsymbol{\Phi}(s)\mathbf{B}$$

The open loop transfer function when the control law is applied is $G_c = \mathbf{K}_c\boldsymbol{\Phi}(s)\mathbf{B} = \mathbf{B}^T\boldsymbol{\Phi}^T(s)\mathbf{K}_c^T$, so that the above expression can be written:

$$G_c^T(-s) + G_c(s) = \mathbf{B}^T\boldsymbol{\Phi}^T(-s)\mathbf{Q}_c\boldsymbol{\Phi}(s)\mathbf{B} - G_c^T(-s)G_c(s) \tag{6.42}$$

which can be rewritten:

$$[1 + G_c^T(-s)][1 + G_c(s)] = 1 + \mathbf{B}^T\boldsymbol{\Phi}^T(-s)\mathbf{Q}_c\boldsymbol{\Phi}(s)\mathbf{B} \tag{6.43}$$

If we define the *return difference function*: $F_c(s) \equiv 1 + G_c(s)$. The second member of (6.43) can be factorized as:

$$1 + \mathbf{B}^T\boldsymbol{\Phi}^T(-s)\mathbf{Q}_c\boldsymbol{\Phi}(s)\mathbf{B} = \Delta_c(s)\Delta_c(-s) \tag{6.44}$$

equation (6.43) can be rewritten:

$$F_c(s)F_c(-s) = \Delta_c(s)\Delta_c(-s)$$

and so:

$$F_c(s) = \Delta_c(s)$$

leading the expression representing the open loop transfer function, with state vector feedback:

$$G_c(s) = \Delta_c(s) - 1$$

It can be observed that the *Riccati equation* has been solved through factorization, determining $\Delta_c(s)$. In fact, $G_c(s)$ has been obtained, which is equivalent to determine $\mathbf{K}_c$, because both are related by expression $G_c(s) = \mathbf{K}_c\boldsymbol{\Phi}\mathbf{B}$. That is, the solution of the LQR problem can be obtained by only manipulating transfer functions, [58], [76].

Once the above expressions have been obtained, the affirmation previously made about robustness of LQR controllers can be demonstrated.

If $\mathbf{Q}_c$ is factorized, $\mathbf{Q}_c = \mathbf{H}_1^T\mathbf{H}_1$ and substituting $s = j\omega$ in the *Riccati equation* (6.43) we get:

$$\|1 + G_c(j\omega)\|^2 = 1 + \|\mathbf{H}_1\boldsymbol{\Phi}(j\omega)\mathbf{B}\|^2$$

and so:

$$\|1 + G_c(j\omega)\| > 1 \tag{6.45}$$

If this condition is interpreted in the polar plane, the plot of $G_c(j\omega)$ cannot be within a circle with center $(-1, 0)$ and radius 1, so that a *phase margin* greater than *60 degrees* and a *gain margin* between 1/2 and ∞ is assured. An analogous development for the observation problem leads to the same results.

Calculation procedure

From the previous results we can summarize that the LQG problem can be solved by the manipulation of transfer functions. The procedure would be to specify matrix $\mathbf{Q}_c$ in a factorized representation as $\mathbf{Q}_c = \mathbf{H}_1^T\mathbf{H}_1$, which is equivalent to specifying the desired transfer function $G_{dc}(s) = \mathbf{H}_1\boldsymbol{\Phi}(s)\mathbf{B}$. From this transfer function and by using the factorization of the *return difference* equation:

$$[1 + G_c^T(-s)][1 + G_c(s)] = 1 + G_{dc}^T(-s)G_{dc}(s) \tag{6.46}$$

the open loop transfer function of LQR $G_c(s)$ can be obtained.

Following the same procedure, by specifying matrix $\mathbf{Q}_o$ in a factorized representation as $\mathbf{Q}_o = \mathbf{H}_2\mathbf{H}_2^T$ or by specifying the transfer function $G_{do}(s) = \mathbf{C}\boldsymbol{\Phi}(s)\mathbf{H}_2$, and by using the factorized *return difference* equation:

$$[1 + G_o(s)][1 + G_o^T(-s)] = 1 + G_{do}(s)G_{do}^T(-s) \tag{6.47}$$

$G_o(s)$ can be obtained.

With these transfer functions, both the gains of the control law $\mathbf{K}_c$ and those of the *Kalman filter* $\mathbf{K}_o$ can be calculated.

6.2.4 Loop transfer recovery (LTR)

In the previous section it has been shown that LQR regulators have excellent robustness conditions, but they can only be used when direct access to the state is available and neither observer or *Kalman filter* is included. As has been shown previously, the introduction of the *Kalman filter*, although not modifying the closed loop transfer function, modifies the open loop one and also the stability margins.

To solve the problems presented by the deterioration of robustness caused by the introduction of the observer the method known as LQG/LTR was proposed by Doyle and Stein [42], [43] . The idea put forward is to try to recuperate the open loop transfer function which is provided by the application of the control law alone, because in this way stability is assured, there is little sensitivity, and the temporary specifications are fulfilled. This can be achieved, in theory, acting on the parameters of the *Kalman filter* so that the open loop transfer function approximates the original open loop transfer function (G_c, equation 6.34) when the *Kalman filter* has been introduced (G_{LQG}, equation 6.39).

Properties of the open loop transfer function of control systems based in observers

Let's suppose a control system with state feedback and the corresponding system with observer, as shown in Figures 6.26 and 6.27 respectively. Between the corresponding schema [43], the following properties are fulfilled:

1. The closed loop transfer function is the same in both cases.
2. The open loop transfer function (opening the scheme in Figure 6.27 at point (1)) is the same:
   – In Figure 6.26, $u = -\mathbf{K}_c \mathbf{x} = -\mathbf{K}_c \boldsymbol{\Phi} \mathbf{B}\, u'$
   – In Figure 6.27, $u = -\mathbf{K}_c \hat{\mathbf{x}} = -\mathbf{K}_c \boldsymbol{\Phi} \mathbf{B}\, u'$
   due to the fact that $\hat{\mathbf{x}} = \boldsymbol{\Phi}\mathbf{B}u'$, because both systems (plant and observer) have the same input signal.
3. The transfer functions opening the complete scheme (6.27) in (2) are different.
   The difference in this case is due to the fact that the dynamics of the observation error $(y - \hat{y})$ are excited, if the loop is opened (or if disturbances are injected) in (2), a fact that does not occur if the loop is opened in (1) (6.27).

Loop transfer recovery conditions

Loop transfer recovery can be achieved by making the *Kalman filter* gains depend on a determined parameter q. Then, we have:

$$\lim_{q\to\infty} \frac{\mathbf{K}_o(q)}{q} = \mathbf{BW} \tag{6.48}$$

with $\mathbf{W}$ being any non singular matrix. The open loop LQG asymptotically approximates LQR.

The open loop transfer function with observer is given by:

$$G_{LQG} = \mathbf{K}_c(s\mathbf{I} - \mathbf{A} + \mathbf{BK}_c + \mathbf{K}_o\mathbf{C})^{-1}\mathbf{K}_o \ \ \mathbf{C}\boldsymbol{\Phi}\mathbf{B} \tag{6.49}$$

If $\boldsymbol{\Psi} = (s\mathbf{I} - \mathbf{A} + \mathbf{BK}_c)^{-1}$ and taking into account the matrix inversion lemma,

$$(\mathbf{A} + \mathbf{BCD})^{-1} = \mathbf{A}^{-1} - \mathbf{A}^{-1}\mathbf{B}(\mathbf{C}^{-1} + \mathbf{DA}^{-1}\mathbf{B})^{-1}\mathbf{DA}^{-1} \tag{6.50}$$

this expression can be written as:

$$\begin{aligned} G_{LQG} &= \mathbf{K}_c[\boldsymbol{\Psi} - \boldsymbol{\Psi}\mathbf{K}_o(1 + \mathbf{C}\boldsymbol{\Psi}\mathbf{K}_o)^{-1}\mathbf{C}\boldsymbol{\Psi}]\mathbf{K}_o \ \ \mathbf{C}\boldsymbol{\Phi}\mathbf{B} = \\ &= [\mathbf{K}_c\boldsymbol{\Psi}\mathbf{K}_o - \mathbf{K}_c\boldsymbol{\Psi}\mathbf{K}_o(1 + \mathbf{C}\boldsymbol{\Psi}\mathbf{K}_o)^{-1}\mathbf{C}\boldsymbol{\Psi}\mathbf{K}_o] \ \ \mathbf{C}\boldsymbol{\Phi}\mathbf{B} = \\ &= [\mathbf{K}_c\boldsymbol{\Psi}\mathbf{K}_o(1 - (1 + \mathbf{C}\boldsymbol{\Psi}\mathbf{K}_o)^{-1})\mathbf{C}\boldsymbol{\Psi}\mathbf{K}_o] \ \ \mathbf{C}\boldsymbol{\Phi}\mathbf{B} = \\ &= \mathbf{K}_c\boldsymbol{\Psi}\mathbf{K}_o(1 + \mathbf{C}\boldsymbol{\Psi}\mathbf{K}_o)^{-1} \ \ \mathbf{C}\boldsymbol{\Phi}\mathbf{B} \end{aligned} \tag{6.51}$$

If in this expression the loop transfer recovery condition given in equation (6.48) is applied:

$$G_{LQG} = \mathbf{K}_c\boldsymbol{\Psi}q\mathbf{BW}(1 + \mathbf{C}\boldsymbol{\Psi}q\mathbf{BW})^{-1} \ \ \mathbf{C}\boldsymbol{\Phi}\mathbf{B}$$

if $q \to \infty$:

$$G_{LQG} = \mathbf{K}_c\boldsymbol{\Psi}\mathbf{BW}(\mathbf{C}\boldsymbol{\Psi}\mathbf{BW})^{-1} \ \ \mathbf{C}\boldsymbol{\Phi}\mathbf{B}$$

By applying once more the matrix inversion lemma (6.50) to this expression and substituting $\boldsymbol{\Psi}$ by its value we get:

$$\begin{aligned} G_{LQG} &= \mathbf{K}_c\boldsymbol{\Psi}\mathbf{BW}(\mathbf{C}\boldsymbol{\Psi}\mathbf{BW})^{-1} \ \ \mathbf{C}\boldsymbol{\Phi}\mathbf{B} = \\ &= \mathbf{K}_c\boldsymbol{\Phi}\mathbf{BW}(1 + \mathbf{K}_c\boldsymbol{\Phi}\mathbf{B})^{-1}[\mathbf{C}\boldsymbol{\Phi}\mathbf{BW}(1 + \mathbf{K}_c\boldsymbol{\Phi}\mathbf{B})^{-1}]^{-1} \ \ \mathbf{C}\boldsymbol{\Phi}\mathbf{B} = \\ &= \mathbf{K}_c\boldsymbol{\Phi}\mathbf{BW}(\mathbf{C}\boldsymbol{\Phi}\mathbf{BW})^{-1} \ \mathbf{C}\boldsymbol{\Phi}\mathbf{B} = \\ &= \mathbf{K}_c\boldsymbol{\Phi}\mathbf{B} \end{aligned}$$

The final expression obtained is:

$$G_{LQG} = \mathbf{K}_c\boldsymbol{\Phi}\mathbf{B}$$

which is identical to expression (6.34) corresponding to G_c. The convergence of the open loop transfer functions when $q \to \infty$ has been demonstrated, when coping with a structure with direct state feedback or with observer (*Kalman filter*).

Loop transfer recovery method

In order for $\mathbf{K}_o$ to fulfill the *recuperation condition* the *Kalman filter* must be modified. To do this the *Kalman filter* is designed with some fictitious covariance matrices. The following will be used:

$$\mathbf{Q} = \mathbf{Q}_o + q^2\mathbf{B}\mathbf{V}\mathbf{B}^T \tag{6.52}$$

$$\mathbf{R} = \mathbf{R}_o \tag{6.53}$$

with $\mathbf{V}$ being any non singular matrix and where $\mathbf{Q}_o$ and $\mathbf{R}_o$ are nominal covariance matrices and q a parameter.

The *Kalman filter* is calculated from the modified covariance matrix:

$$\mathbf{K}_o = \mathbf{P}\mathbf{C}^T\mathbf{R}^{-1} \tag{6.54}$$

$$\mathbf{A}\mathbf{P} + \mathbf{P}\mathbf{A}^T + \mathbf{Q} - \mathbf{P}\mathbf{C}^T\mathbf{R}^{-1}\mathbf{C}\mathbf{P} = \mathbf{0} \tag{6.55}$$

Introducing previous covariance matrices in the *Riccati equation* (6.55), we get:

$$\mathbf{A}\mathbf{P} + \mathbf{P}\mathbf{A}^T + \mathbf{Q}_o + q^2\mathbf{B}\mathbf{V}\mathbf{B}^T - \mathbf{P}\mathbf{C}^T\mathbf{R}^{-1}\mathbf{C}\mathbf{P} = \mathbf{0} \tag{6.56}$$

and by dividing the expression by q^2 we get,

$$\mathbf{A}\frac{\mathbf{P}_o}{q^2} + \frac{\mathbf{P}_o}{q^2}\mathbf{A}^T + \frac{\mathbf{Q}_o}{q^2} + \mathbf{B}\mathbf{V}\mathbf{B}^T - q^2(\frac{\mathbf{P}}{q^2})\mathbf{C}^T\mathbf{R}^{-1}\mathbf{C}(\frac{\mathbf{P}}{q^2}) = \mathbf{0} \tag{6.57}$$

Notice that in equation (6.57), there exist two kinds of variables: q, that is the parameter whose value is modified for loop transfer recovery purposes, and matrix $\mathbf{P}$, which condition the values of the *Kalman filter* gains, and that can only be known once the value of q has been fixed. So, if $q \to \infty$ we get:

$$q^2(\frac{\mathbf{P}}{q^2})\mathbf{C}^T\mathbf{R}^{-1}\mathbf{C}(\frac{\mathbf{P}}{q^2}) \xrightarrow{q\to\infty} \mathbf{B}\mathbf{V}\mathbf{B}^T$$

and taking into account the value of $\mathbf{K}_o$ (equation 6.54):

$$\frac{\mathbf{K}_o(q)\mathbf{R}\mathbf{K}_o^T(q)}{q^2} \xrightarrow{q\to\infty} \mathbf{B}\mathbf{V}\mathbf{B}^T$$

can be obtained which when broken down adequately leads to:

$$\frac{\mathbf{K}_o(q)}{q} \xrightarrow{q\to\infty} \mathbf{B}\underbrace{\mathbf{V}^{\frac{1}{2}}(\mathbf{R}^{\frac{1}{2}})^{-1}}_{\mathbf{W}} = \mathbf{B}\mathbf{W}$$

So, the *recovery condition* is fulfilled (equation 6.48).

6.2.5 LQG/LTR design method

In this section, the necessary steps to calculate $G_R(s)$ of a monovariable control loop are explained. The normalized structure is shown in Figure 6.29, where:

$G_p(s)$ is a minimum phase plant,
$w(s)$ is a white random noise at the input,
$v(s)$ is a white random noise at the output and
$G_R(s)$ is the regulator.

The plant model has the following internal description:

$$\begin{aligned} \dot{\mathbf{x}} &= \mathbf{Ax} + \mathbf{Bu} + \mathbf{w} \\ y &= \mathbf{Cx} + \mathbf{v} \end{aligned} \tag{6.58}$$

with the pair $(\mathbf{A}, \mathbf{B})$ controllable and the pair $(\mathbf{A}, \mathbf{C})$ observable. The objective is to determine $G_R(s)$ so that:

- The poles of $1 + G_R(s)G_p(s)$ have negative real part (stable closed loop).
- The perturbed system is stable (in spite of measurement errors and modelling errors).
- Some particular conditions which will appear as constraints for the resulting open loop $G_R(s)G_p(s)$ are fulfilled.

The design method of $G_R(s)$ is constituted by three steps:

1. Determine the *control law* by state vector feedback, that is, calculate $\mathbf{K}_c$ so that the open loop transfer function

$$G_c(s) = \mathbf{K}_c\boldsymbol{\Phi}(s)\mathbf{B}$$

   fulfills the pre-established specifications in the frequency domain. The controllability of $(\mathbf{A}, \mathbf{B})$ guarantees the existence of $\mathbf{K}_c$. The function $G_c(s)$ obtained in this way, besides assuring the fulfillment of the particular specifications of the problem, will have a great degree of robustness, having wide *stability margins*. In this way, the first two conditions are covered. This control structure will not usually be realizable, because the state variables will not often be accessible.
2. Determine an observer so that compensation will be realizable. The observer structure will be that of the *Kalman filter* for the non deterministic case. The combination of both the control law and the *Kalman filter* will give rise to a regulator with transfer function:

$$G_R(s) = \mathbf{K}_c(s\mathbf{I} - \mathbf{A} + \mathbf{BK}_c + \mathbf{K}_o\mathbf{C})^{-1}\mathbf{K}_o \tag{6.59}$$

   This regulator can be obtained, as mentioned in previous sections, by any usual state space method or by manipulating transfer functions. Particularly, expression (6.59) can also be calculated by resolving a polynomial equation [104].

3. Determine the robustness of the system with a G_R regulator. If it is not adequate, the gains of the *Kalman filter* ($\mathbf{K}_o$) are recalculated with smaller covariance of the measurement noise (or greater covariance of the process noise), in order to satisfy the *recovery condition*. In this way $G_R(s)G_p(s)$ will asymptotically approximate $G_c(s)$. That is:

$$G_R(j\omega)G_p(j\omega) \approx G_c(j\omega) \tag{6.60}$$

in a relevant frequency range to fulfill *robustness* and *performance*. In fact, at high frequencies $G_c(j\omega)$ will differ substantially from $G_R(j\omega)G_p(j\omega)$, because the first has a slope of $-20\ dB/dec$ while the second has a slope of $-40\ dB/dec$. These characteristics, however, allow the LQG/LTR to introduce additional *robustness* to the system.

6.2.6 Output recovery

The method explained in the previous section is known as LQG/LTR with input recovery, because the loop breaking point to calculate the open loop transfer function is situated at the plant input.

A dual procedure (LQG/LTR with output recovery) exists in which the loop breaking point is situated at the output of the plant. The method is useful when uncertainties at the output of the plant can be studied or when the system has more inputs than outputs.

The method is based on the recuperation of the open loop transfer function of the *Kalman filter*, due to the fact that the duality between the regulation and observation problems allows similar robustness characteristics to be obtained. The matrices corresponding to the performance indexes $\mathbf{Q}_c$ and $\mathbf{R}_c$ are modified in this case. The problem is that this procedure modifies the location of the poles of the global system (not those of the observer, as with input recovery), and the degrees of freedom to achieve specifications are diminished. Particularly, at the limit, the closed loop poles will be near to plant zeros, leading to almost unacceptable behaviour.

The method is summarized in the following steps:

1. Solve the LQR problem, by adequately selecting $\mathbf{Q}_c$ and $\mathbf{R}_c$.
2. Calculate the *Kalman filter.*
3. Determine the robustness of the resulting system. If it is not adequate, it is necessary to recover the transfer function repeating step 1 with modified weighting matrices:

$$\begin{aligned}\mathbf{Q}_c &= \mathbf{Q}_c^o + q^2\mathbf{C}^T\mathbf{V}\mathbf{C}\\ \mathbf{R}_c &= \mathbf{R}_c^o\end{aligned}$$

with $\mathbf{V}$ being any non singular matrix. This process is repeated until an adequate value of q is found.

6.2.7 Application to the distributed solar collector field

This section presents a control structure in which the controller is an LQG/LTR developed to obtain a robust fixed parameters controller, able to acceptably control the distributed solar collector field under a wide range of operating conditions [102].

Control structure

In the development of the robust controller, low order models such as those proposed in chapter 2 have been used to show that, even in those cases in which the modelling uncertainty is high, the proposed robust control scheme leads to acceptable results with a low order controller. The feedforward controller described in chapter 3 has also been used in this case, in series with the LQG/LTR robust controller, to compensate for disturbances acting on the system. The global control structure is shown in Figure 6.30.

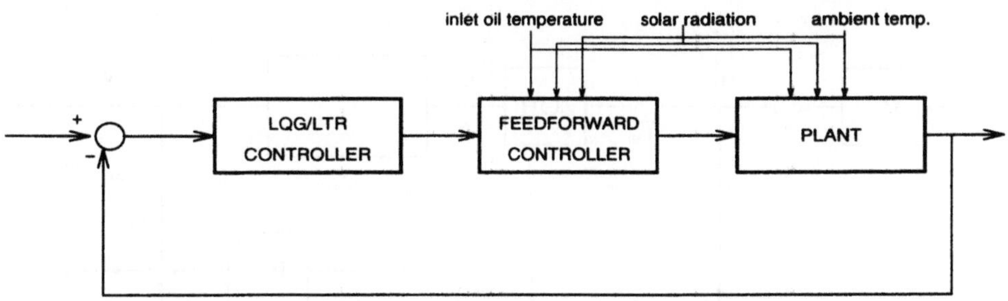

Fig. 6.30. LQG/LTR control scheme with feedforward controller

The LQR and LQG regulators in discrete time domain

Given the plant model:

$$\begin{aligned} \mathbf{x}(k+1) &= \mathbf{A}\mathbf{x}(k) + \mathbf{B}\mathbf{u}(k) \\ \mathbf{y}(k) &= \mathbf{C}\mathbf{x}(k) \end{aligned}$$

The purpose of the LQR regulator is to determine the control signal $\mathbf{u}$ in order to minimize the following functional:

$$\mathcal{J} = \mathbf{x}_N^T \mathbf{H}_c \mathbf{x}_N + \sum_{k=0}^{N-1} \left(\mathbf{x}^T(k) \mathbf{Q}_c \mathbf{x}(k) + \mathbf{u}^T(k) \mathbf{R}_c \mathbf{u}(k) \right)$$

$$\text{with:} \quad \mathbf{Q}_c = \mathbf{Q}_c^T \geq 0, \qquad \mathbf{R}_c = \mathbf{R}_c^T \geq 0$$

$\mathbf{x}$ and $\mathbf{x}_N$ being the state and the final state vectors respectively. The LQG problem is based on the separation theorem, according to which the control strategy can be divided into two parts:

1. An optimal control problem, from which the regulation by feedback of state variables is obtained.

$$\mathbf{u} = -\mathbf{K}_c\hat{\mathbf{x}} \tag{6.61}$$

where $\mathbf{K}_c$ is found by iteration of a *Riccati equation.*

2. An optimal filtering problem which solves the observation using the *Kalman filter*, whose corresponding dynamics is given by the system:

$$\begin{aligned}\hat{\mathbf{x}}(k+1) \;=\; & \mathbf{A}\hat{\mathbf{x}}(k) + \mathbf{B}\mathbf{u}(k) + \mathbf{K}_o(k+1)\ [\mathbf{y}(k+1) - \\ & \mathbf{C}(\mathbf{A}\hat{\mathbf{x}}(k) + \mathbf{B}\mathbf{u}(k))]\end{aligned} \tag{6.62}$$

where $\mathbf{K}_o$ is obtained from a *Riccati equation* dual to the previous one.

The resulting structure is shown in Fig. 6.31.

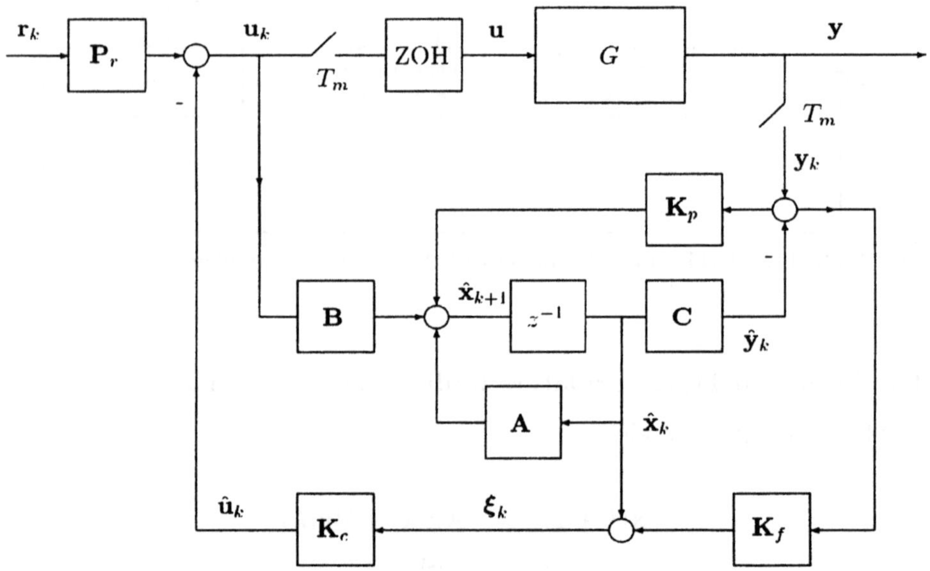

Fig. 6.31. Block diagram of the LQG controller

Loop Transfer Recovery (LTR)

Asymptotic recovery techniques have been developed for continuous time, minimum phase systems [42]. It is, therefore, desirable to have a similar procedure for discrete time systems, even though the LQR discrete controller does not have the same properties as are present in the continuous case, such as stability margins.

The solution of the controller LQR is a linear feedback of the state vector and it gives rise to the open loop transfer function:

$$G_{LQR} = \mathbf{K}_c \boldsymbol{\Phi} \mathbf{B} \qquad \text{with} \qquad \boldsymbol{\Phi} = (z\mathbf{I} - \mathbf{A})^{-1}$$

By mathematical manipulation in Fig. 6.31, the transfer function of the open loop $G_{LQG}(z)$, can be obtained and this depends on the parameters of the *control law* and of the *Kalman filter*, which in turn depend on the *costs* assigned to solving the control objective and covariances of observation problems. It is a well known fact that the closed loop transfer function is the same as for the LQR problem, thus the closed loop temporary specifications solved by the application of the control law will continue to be fulfilled. However, the open loop transfer function becomes:

$$G_{LQG} = [\mathbf{K}_c(\mathbf{I}-\mathbf{K}_o\mathbf{C})[z\mathbf{I}-(\mathbf{A}-\mathbf{B}\mathbf{K}_c)(\mathbf{I}-\mathbf{K}_o\mathbf{C})]^{-1}(\mathbf{A}-\mathbf{B}\mathbf{K}_c)\mathbf{K}_o+\mathbf{K}_c\mathbf{K}_o]\ \mathbf{C}\boldsymbol{\Phi}\mathbf{B}$$

With this situation the stability margins have changed, and in most cases this variation usually deteriorates the phase margin of the compensated system, causing the consequent loss of robustness.

In the continuous time case, the LQG/LTR method solves the problem in the case of minimum-phase systems, in the following way:

$$\begin{aligned} \mathbf{Q}_o &= \mathbf{Q}_o^0 + q^2\mathbf{B}\mathbf{B}^T \\ \mathbf{R}_o &= \mathbf{R}_o^0 \end{aligned}$$

As $q^2 \to \infty$, the open loop transfer function of the LQG problem approaches the LQR one.

In the case of discrete systems [77], it can be shown that for minimum phase systems such as $det(\mathbf{CB}) \neq 0$ and using the filtered version of the *Kalman filter* as observer, perfect recovery is obtained acting on the parameters of the LQR. In any case, when these conditions are not fulfilled, the same procedure usually leads to improvements of robustness.

Controller design

The effect of loop transfer recovery on the increase in robustness of the controlled system has been analyzed first. The controlled system could be expected to be less sensitive to changing dynamics due to set point variations or unmodelled dynamics.

In order to carry out the design of the LQG/LTR controller a linear model must first be considered. In this case, the reduced order model considered in chapter 3 has been employed.

In order to show the advantages of the LTR method, the controller has been designed using a model of the solar power plant (Type A Model): $G(z) = 0.1z^2/(1-0.9z^{-1})$, which corresponds to high flow operating conditions. The model has been augmented with an integrator in order to avoid static errors and the controllable canonical form has been used. The weighting matrices of the LQ problem have been used equal to:

$$\mathbf{Q}_c = \mathbf{C}^T\mathbf{C}; \quad \mathbf{R}_c = 17$$

where $\mathbf{C}$ is, as usual, the output matrix of the internal description of the plant. The value $\mathbf{R}_c = 17$ has been chosen from simulation studies in order to achieve good tracking characteristics.

Since it is not possible to obtain an estimation of the noises corresponding to the *Kalman filter* (amongst other facts, the internal description used has no physical meaning), the covariance matrices of the *Kalman filter* have been chosen arbitrarily. To be precise, a LTR-i form has been used:

$$\mathbf{Q}_o = \mathbf{B}\mathbf{B}^T; \quad \mathbf{R}_o = \rho$$

where $\mathbf{B}$ is, as usual, the input matrix of the internal description of the plant and ρ is a design parameter. According to the LQG/LTR theory, decreases of ρ will probably result in robustness increases.

The behaviour of the simplified linear model $G(z)$ with a design value of $\rho = 0.1$ has been tested both at the design point (maximum flow) and at the extreme point (minimum flow, which corresponds to the same transfer function augmented with z^{-1}). Simplifying assumptions have been made, because fixed values of b_0, b_1 and a have been considered in the reduced order model shown in chapter 3. As can be seen in Figure 6.32, the behaviour of the controlled system is influenced by changes at operating point. If a value of $\rho = 10^{-10}$ is chosen in the design, the controlled system is less sensitive to changes at the operating point (same figure), as is to be expected.

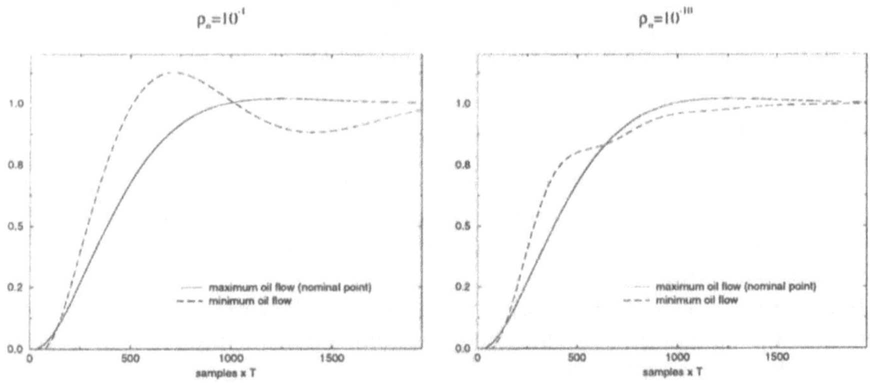

simulation with $\rho = 0.1$ simulation with $\rho = 1e-10$

Fig. 6.32. Robustness increase with loop transfer recovery

Simulation studies

In order to develop a LQG/LTR controller for the distributed solar collector field, a model corresponding to an identification of the step response of the system under a precise operating condition (oil flow equal to 8 l/s) has been chosen. This model (Type B Model) would be very different under other conditions, but, due to the robustness of the controller, stability and good set point tracking is obtained under different operating conditions. The model proves to be equal to:

$$G_1(z) = \frac{0.027z + 0.052}{z^3 - 0.912z^2}$$

As mentioned previously, the model is augmented with an integrator and the controllable canonical form is used as an internal description of $G(z) = \frac{z}{z-1}G_1(z)$. The weighting matrices of the LQ problem stated above have been used.

In Figure 6.33 the Nyquist diagram of the model used for design purposes with the LQG controller is represented, with different values of ρ. It can be seen that, as ρ decreases, the phase margin of the controlled system increases (or any other robustness measure).

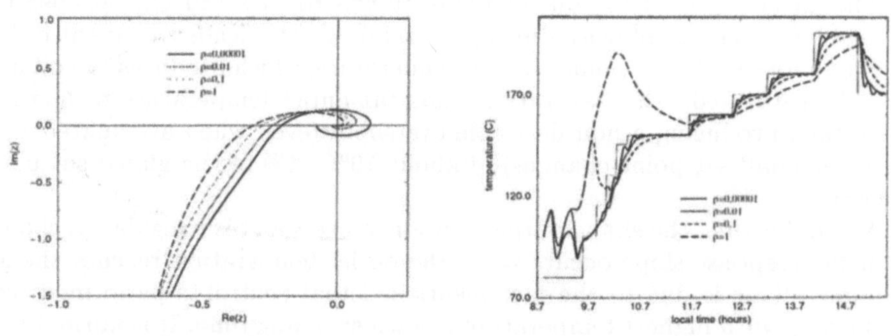

Nyquist plots simulations

Fig. 6.33. Nyquist diagram and simulation with the LQG/LTR controller for different values of ρ

In this case, as ρ diminishes (augmenting the margin phase of the controlled system), faster responses are obtained in spite of set point changes. When coping with input recovery, the transfer function of the closed loop system after the recovery has been performed is expected to be equal to that of the system without recovery, and so, the behaviour of the controlled system. Nevertheless, this affirmation can be made when designing linear time invariant systems, but not when coping with nonlinear systems, such as the one treated here (see Figure 6.33). If recovery should be performed at the output, as $\mathbf{R}_c$ is decreased (increasing the robustness of the controlled system), a faster response of the controlled system will be produced, because the control effort is being less penalized.

The value of $\rho = 0.00001$ has been chosen in simulation of Figure 6.33 to obtain a good trade–off between robustness and the desired shape of the step response. Results of simulations are good compared to other controller approaches including adaptive schemes. In this case, the lack of adaptive behaviour is compensated for by the high robustness of the controller.

Plant results

As with other control approaches, the proposed control scheme has been tested at the solar power plant. First of all, it is important to point out that with the design of the LQG/LTR controller, acceptable results were obtained right from the first test performed at the plant. This fact can be observed in figures showing the first two tests. Figure 6.34 corresponds to a step response test on a clear day (June 8, 1995). Set point, inlet and outlet oil temperatures, direct solar radiation and oil flow are represented). As can be seen, an initial transient exists after which the outlet temperature reaches the first operating point. This transient is mainly due to the following facts:

- The initial value of the oil temperature was too low (40°C), because the field had been out of work during several days. This leads to a great difference between the set point and oil temperature, which produces the oil flow to be saturated. This saturation leads the outlet temperature to increase quickly, producing a non desirable overshoot (even when attenuated by a set of small set point changes) of about 10°C (5% of the global set point change).
- As can be seen, the shape of the mentioned overshoot is irregular (a change in the response slope occurs when the outlet temperature reaches the set point). This is due to the fact mentioned that control is performed over the loop with highest temperature in each sampling time. It is normal that some of the modules which conform a loop should take a long time to reach track position, and even, in certain situations, some adjustments have to be manually made by the field supervisors. When this problem is solved, it is usual that the loop affected will achieve, after a transient, the highest outlet temperature, producing a change in the loop being controlled, with a transient which produces the response to have the mentioned irregular shape (see Figure 6.34).
- In the starting phase of the operation, solar radiation experiments increase due to its daily cycle.

– A smooth increase in the inlet temperature also exists, due to the recirculation of the oil through the field until its temperature is adequate to enter the top of the storage tank.

After this initial transient, good tracking characteristics are obtained, with small control effort and acceptable rise times (about 8 minutes).

Figure 6.35 (set point, inlet and outlet temperature, direct solar radiation and oil flow) shows the result of a test (13/06/95) covering a wide range of oil flows (from 1.8 to 9.8 l/s). As can be seen, after the initial transient, whose causes have been commented, good set point tracking with null steady state error is obtained. The rise time is again of about 8 minutes, with an overshoot less than 1°C when set point changes of 10°C are performed. At the middle of the operation (about 13.5 decimal local time), a large cloud covered the field for 20 minutes, causing the oil flow to be saturated (minimum oil flow) to avoid a drop in the outlet temperature. Under such circumstances, the field is very difficult to control. The LQG/LTR allows the previous set point after the transient in which, antiresonance modes are excited to be recovered. The observed control action was too damped, and so, the controller was slightly detuned.

Results obtained when performing the change can be observed in Figure 6.36 (outlet oil temperature, inlet oil temperature, set point temperature, solar radiation and oil flow respectively) corresponds to a step response test (19/09/95) covering a wide range of oil flow conditions (from 1.8 to 6 l/s). As can be seen, good set point tracking and disturbance rejection was obtained with null steady state error (due to the fact that the controller is implemented including integral action. The rise time was of about 8 minutes without overshoot in a change of 10 degrees at the set point.

Disturbance rejection

It has been observed that, with high degrees of recovery ($\rho \rightarrow 0$), a fast controlled system is obtained when coping with set point changes but, its response to disturbances acting on the system is too vigorous, exciting high frequency dynamics. In order to obtain smoother transients, the design value of $\mathbf{R}_c$ can be increased. In this particular case, it has been observed that this effect can be obtained by slightly decreasing the degree of recovery in the open loop transfer function. As a result of this small change, results obtained with the modified controller are shown in Figure 6.36, in a test with high scattered clouds. As can be seen, transients are smoother and overshoot is diminished with the cost of a small increase in rise time (about 9 minutes). Figure 6.37 shows a zoom corresponding to normal operation to appreciate the tracking characteristics of this controller.

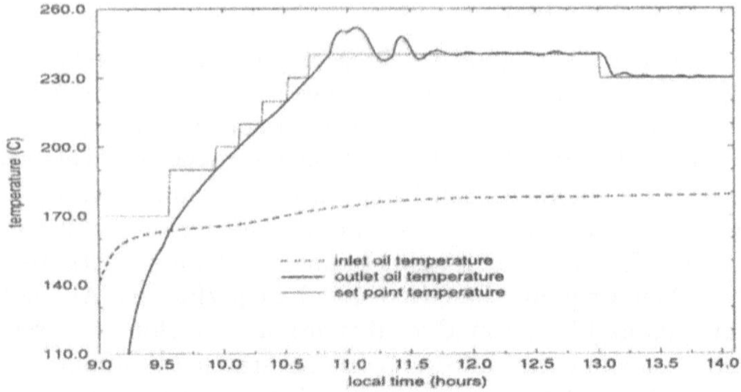

set point, inlet and outlet oil temperature (°C)

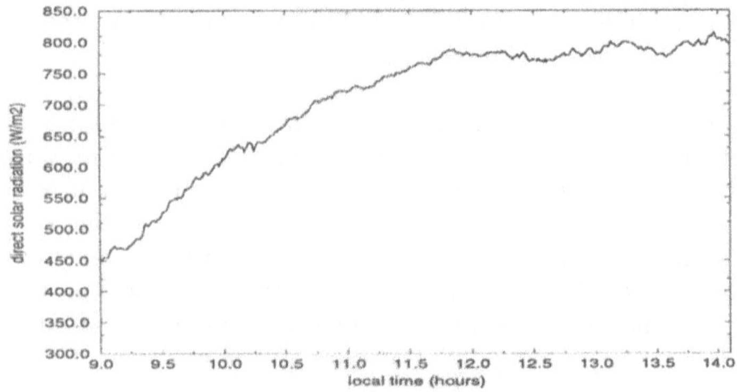

direct solar radiation (W/m^2)

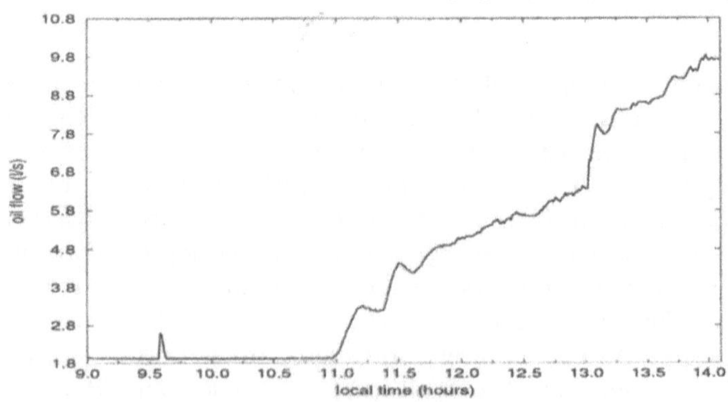

oil flow (l/s)

Fig. 6.34. Test with the LQG/LTR controller (08/06/95)

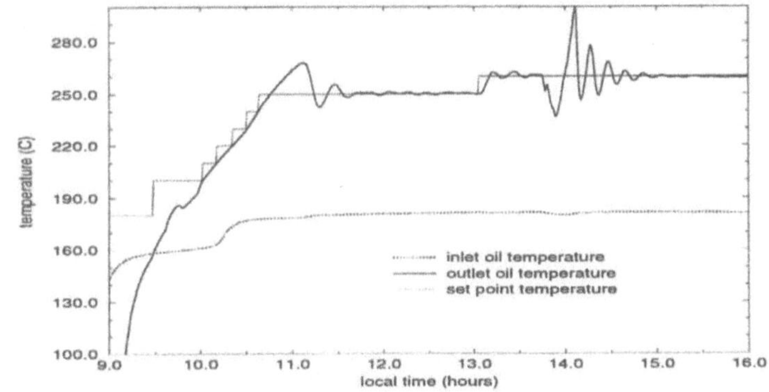

set point, inlet and outlet oil temperature (°C)

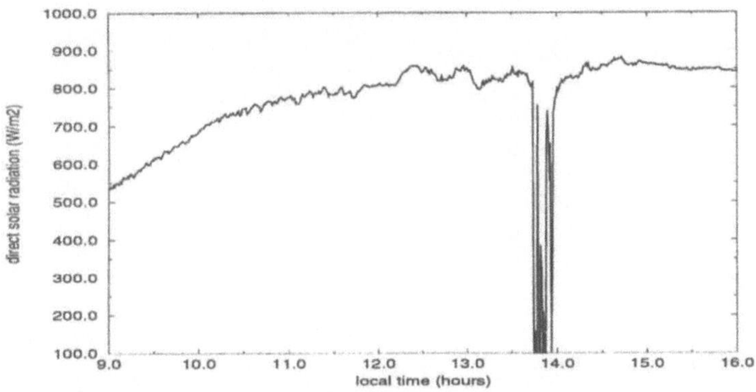

solar radiation (W/m^2)

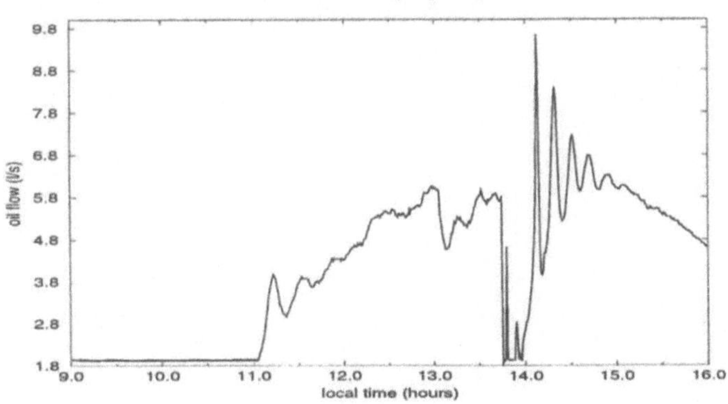

oil flow (l/s)

Fig. 6.35. Test with the LQG/LTR controller (13/06/95)

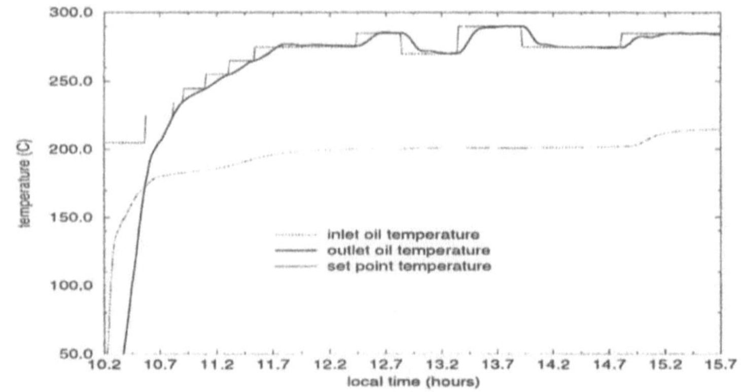

set point, inlet and outlet oil temperature (°C)

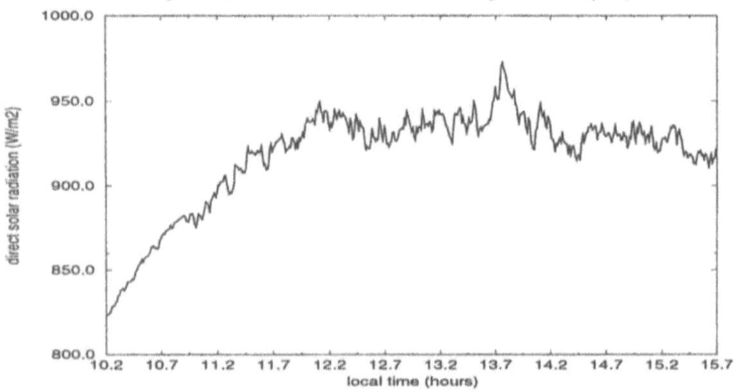

direct solar radiation (W/m^2)

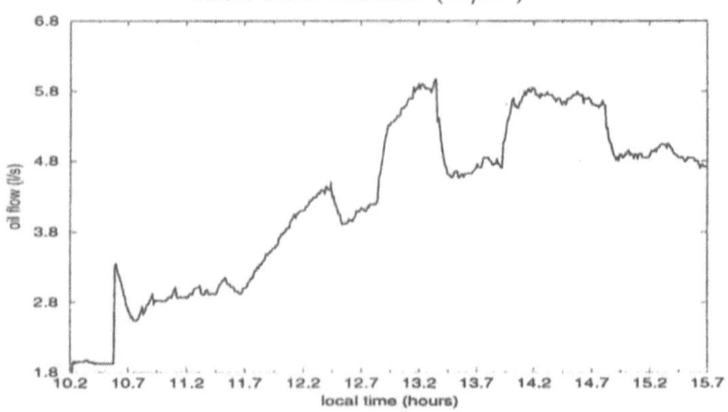

oil flow (l/s)

Fig. 6.36. Test with the LQG/LTR controller (19/09/95)

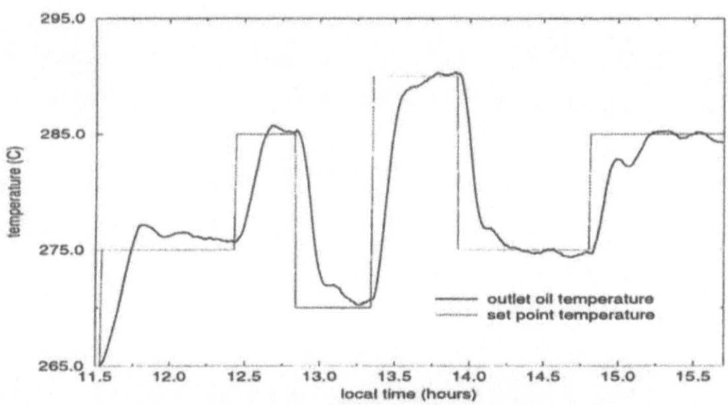

set point and outlet oil temperature (°C)

Fig. 6.37. Test with the LQG/LTR controller (19/09/95)

CHAPTER 7
HEURISTIC FUZZY LOGIC CONTROL

Since Mamdani [79] published his experiences using a fuzzy logic controller on a test-bed plant of a laboratory, many different control approaches have appeared based on this theory and also many applications of this type of controller to a diversity of processes such as warm water plants [60], rotative clinker-cooker furnaces for cement production [69], [61], automatic train operation [123], cargo ship steering [71], robotics [108], [107], and others. An extensive introduction to the historical development, state and concepts involving fuzzy control systems can be found in [72], [73].

A great many advantages of fuzzy logic controllers have been mentioned in literature. One relevant advantage is the possibility of converting a linguistic control strategy based on experience and expert knowledge into an automatic control strategy. Another advantage is that fuzzy logic controllers can easily be implemented. Moreover, it has been demonstrated [115], that by using fuzzy inference systems with triangular partition rule bases with evenly spaced midpoints (TPE systems) like those used in this chapter, the amount of computation required for processing input is independent of the number of rules.

The methodology of fuzzy controllers seems to be the most adequate in cases where the process is difficult to control and there is wide experience in operating the plant, due to the possibility of incorporating such knowledge in terms of qualitative rules. This is the case of the control schema presented in this chapter.

A special subclass of fuzzy inference systems, the TP and TPE systems are used both in the development of an incremental fuzzy PI controller (IFPIC) and in the implementation of a fuzzy logic control (FLC) strategy which provides an adequate control signal under the entire range of possible operating conditions. Both strategies have been implemented in connection with the simple feedforward controller analyzed in chapter 3, which allows for compensating effects of the disturbances acting on the system. Simulation results and actual plant tests are provided.

The first control scheme described is based on an incremental fuzzy PI control structure (IFPIC). Conventional PI control strategies have been commonly used when controlling distributed solar collector fields. Acceptable behaviour can be achieved by using a fixed PI controller designed to work around a determined operating point, but poor performance is often obtained when operating far from the design point because of the changing dynamics of the plant. Adaptive PI control strategies have been applied successfully by means of a low order model approximation of the plant dynamics (chapter 4). Here, other kinds of on line adaptation of PI controller parameters based on experience of operating with the plant is developed. This approach assumes that both nominal controller parameters are available (obtained from some classical tuning methods) and the control structure is a PI one.

The second control scheme (FLC) does not presuppose an explicit controller structure and nominal controller parameters [101]. The control signal is obtained straightforwardly from the fuzzy logic control mechanism and not by calculating adequate controller parameters from which the control signal is obtained. This scheme allows for implementing a nonlinear control strategy.

7.1 Fuzzy logic inference scheme

A fuzzy logic inference scheme is commonly described by a set of fuzzy rules that constitute the control protocol. With these rules, the interconnected relationships between measurable variables and control variables can be expressed.

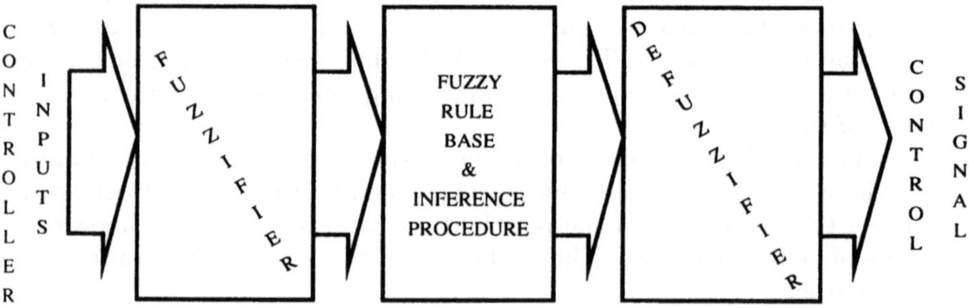

Fig. 7.1. Fuzzy logic inference scheme

In Figure 7.1 three parts can be seen which constitute the design parameters of a fuzzy logic controller: the block fuzzifier; the control block (fuzzy rule base & inference procedure), and the block defuzzifier.

- The fuzzification interface converts the numerical values of the input variables into linguistic variables (fuzzy sets). The conversion requires scale mapping that transforms the range of values of input variables into corresponding universes of discourse.
- The rule based fuzzy control algorithm provides definitions of linguistic control rules which characterize the control policy. In this case, the block

includes decision making logic, which infers fuzzy control actions employing fuzzy implication and the inference rules mentioned.
- The defuzzification block converts the inferred control action, which interpolates between rules that are fired simultaneously, at a continuous signal.

The expert knowledge usually takes the form:

if *(a set of conditions are satisfied)* **then** *(a set of consequences can be inferred)*

The set of conditions (fuzzy sets) belongs to the input domain and the set of consequences to the output domain. The fact that various rules can be fired simultaneously is due to the conversion of values obtained from sensors to linguistic terms, assigning a membership function μ_A to each one. If triangular partition rule bases are used [115], the input and output universes are subdivided using triangular membership functions of the form:

$$\mu_{A_i}(x) = \begin{cases} (x - a_{i-1})/(a_i - a_{i-1}) & \text{if } a_{i-1} \le x \le a_i \\ (-x + a_{i+1})/(a_{i+1} - a_i) & \text{if } a_i \le x \le a_{i+1} \\ 0 & \text{otherwise.} \end{cases} \tag{7.1}$$

being the membership values of $A_i \neq 0$ only at the interval (a_{i-1}, a_{i+1}). Point a_i is the unique element that has membership value 1 at A_i (midpoint of A_i). A triangular decomposition of a universe consists of a sequence of triangular fuzzy subsets $A_1, ..., A_n$ such that the leftmost and rightmost fuzzy regions satisfy $\mu_{A_1} = 1$ and $\mu_{A_n} = 1$ at its midpoints (we are assuming that the triangular decomposition forms a fuzzy partition of the underlying universe, that is: $\sum_{i=1}^{n} \mu_{A_i}(u) = 1$ for every u belonging to the universe). A decomposition of the input and output domains that satisfies these requirements is called a TP (triangular partition) system. The inference process can be made more efficient by requiring the membership functions to be isosceles triangles with bases of the same width. These fuzzy inference systems with evenly spaced midpoints are called TPE systems.

The inference mechanism in a TP or TPE system is very straightforward and efficient. As is demonstrated in [115], considering SISO systems (the extension to MIMO system is straightforward) if A_i and A_{i+1} are two fuzzy sets (with midpoints a_i and a_{i+1} respectively) providing non zero membership for x, the appropriate action for any input at an interval $[a_i, a_{i+1}]$ using weighted averaging defuzzification is given by

$$z = \frac{\mu_{A_i}(x)c_r + \mu_{A_{i+1}}(x)c_s}{\mu_{A_i}(x) + \mu_{A_{i+1}}(x)} = \frac{x(c_s - c_r) + a_{i+1}c_r - a_i c_s}{a_{i+1} - a_i} \tag{7.2}$$

with c_s and c_r being the midpoints of C_s and C_r (fuzzy regions in the output space).

Equation (7.2) shows that only two completely determined constants $(c_s - c_r)/(a_{i+1} - a_i)$ and $(a_{i+1}c_r - a_i c_s)/(a_{i+1} - a_i)$, are required to obtain the control action, allowing a fuzzy inference system in tabular form to be represented. The amount of computation required for processing input is independent of the number of rules with a TPE system. The tabular information can be stored in a manner that permits direct addressing, avoiding searching procedures to find the appropriate "rules" in the inference table.

A fuzzy associative memory (FAM) can be used instead of the table. A FAM is a k-dimensional table where each dimension corresponds to one of the input universes of the rules. The ith dimension of the table is indexed by the fuzzy sets that comprise the decomposition of the ith input domain. The FAM representation may be modified to produce a numeric inference (NI) table. In this way, the indices will represent the corresponding midpoints of the set. The entries in the table are the midpoints of the consequent of the associated rule.

7.2 Incremental fuzzy PI control (IFPIC)

In this section, a control scheme to improve performance of classical PI controllers is used. The approach assumes that nominal controller parameters are available (obtained from some classical tuning methods, as done by Tzafestas and Papanikolopoulos in [120]). By using an appropriate fuzzy matrix, small changes in each controller parameter are performed in order to improve transients and steady state performance of the closed loop system.

As has been mentioned, the possibilities of PI controllers are limited in order to control complex plants, like the distributed solar collector field presented in this book. From experience of operating the field, a fuzzy matrix that contains this knowledge (in condensed rule form) is obtained for each parameter of a PI controller. These matrices describe the changes that the nominal controller parameters experiment.

There have been some methods, such as the one presented by King and Mamdani [61] in which the rule justification is made by referring to a closed system trajectory in a phase plane. Knowledge of parameter adjusting based on phase plane analysis (e.g., overshoot, rise time) and intuition or previous knowledge of the behaviour of the closed loop system are required. As mentioned in [72], a slightly modified method consists of tracking the linguistic trajectory of a closed loop system in a "linguistic phase plane". The main idea is that scale mappings would first be adjusted to approximately yield a desired trajectory behaviour.

In the paper of Tzafestas and Papanikolopoulos [120], three different classical PID tuning techniques are used (Ziegler-Nichols, analytical and Kalman PID tuning techniques). Here, another approximation based on pole-placement techniques is used. Empirical rules are used to build a 7×7 fuzzy matrix for each parameter, which gives an acceptable quantization. The basis for the control rules is the error $tref - y$ (where $tref$ is the set point and y is the outlet oil temperature) and its increment.

In order to obtain good performance, two main rules have to be taken into account:

- To decrease overshoot, the integral term has to be decreased when the output is approaching the set point. To diminish the rise time, the integral term has to be increased during the transient.
- To decrease rise time, the proportional term has to be increased during the transient and decreased when the output is approaching the set point.

These rules and the previous knowledge of the controlled system behaviour produces the control matrix of each parameter. Each matrix, can be described (if triangular partitions are chosen for the input and output universes of discourse) by the midpoint element of each region. The inputs chosen are the error e and its increment Δe. In this way, the proportional and integral terms (controller parameters) can be defined:

$$\begin{aligned} P &= P_{nominal} + \Delta P\{E, \Delta E\}_P \\ I &= I_{nominal} + \Delta P\{E, \Delta E\}_I \end{aligned}$$

where P denotes the proportional term, I denotes the integral term, the subindex *nominal* refers to nominal parameters, ΔP is the value obtained from the matrix (subindex P of proportional term matrix and subindex I of integral term matrix). Notice that another form of designing could be by using only one matrix and by multiplying by scale factors [120].

7.2.1 Application to the distributed solar collector field

The fuzzy logic inference scheme explained in the previous section has been used to control the distributed solar collector field ACUREX.

In order to define input and output signals of the fuzzy logic inference scheme, the series feedforward controller developed in chapter 3 has also been used here, to make an adjustment in the flow input, aimed at eliminating the change in outlet temperature caused by the variations in solar radiation and inlet temperature. Again, the calculation employed is:

$$u_f = \frac{0.7869\ I\ -\ 0.485\ (u - 151.5) - 80.7}{(u\ -\ T_{in})}$$

where u_f is the oil flow, u in this case is the temperature set point given by the fuzzy logic controller, T_{in} is the inlet oil temperature and I is the effective solar radiation. The feedforward is placed in series with the fuzzy logic controller and so, the output signal of the fuzzy controller is the set point temperature for the feedforward term and not the oil flow (this is calculated by the feedforward controller).

The range of parameter variation is obtained from the dynamic study of the plant made in chapter 2 and taking into account the effect of the feedforward term. It is important to clarify that the further development, parallel to that made for the adaptive PI in chapter 4, is used to obtain a conservative range of the controller parameters. This range is finally adjusted using the experience of operating the plant.

Low order approximated models were obtained in chapter 2. Particularly, the discrete transfer function of the Type A model was obtained:

$$G(z^{-1}) = z^{-k}\frac{bz^{-1}}{1-az^{-1}}$$

and at the high flow level condition k = 1. Step responses were obtained covering the range of possible oil flows with different operating conditions (solar radiation and inlet oil temperatures). As has been mentioned, from these observations and from stability analysis, it was found that appropriate parameters that describe the system model in the range of operating points (including the feedforward term) are: $0.8 \leq a \leq 0.95, 0.9 \leq K_{est} \leq 1.2$, where K_{est} is the static gain of the plant, that is, $K_{est} = b/(1-a)$.

The transfer function of a PI controller in the continuous time domain is described by $G_{PI}(s) = K_{PI}(1 + \frac{1}{T_I s})$, where K_{PI} is the analog gain and T_I is the integral time constant. Discretizing the PI transfer function, using the Backward-Euler approximation, one obtains the classical discrete time PI transfer function $G_{PI}(z) = \frac{g_0 + g_1 z^{-1}}{1-z^{-1}}$, where $g_0 = K_{PI}(T_I + T)/T_I$ and $g_1 = -K_{PI}$, being T the sampling time.

If the controller zero is chosen to cancel the plant pole, i.e., choosing $g_1/g_0 = -a$, the system closed loop transfer function is given by

$$G_{cl}(z) = \frac{g_0 b}{z^2 - z + g_0 b}$$

The closed loop characteristic polynomial roots are given by $z_{1,2} = 0.5 \pm 0.5\sqrt{1-4g_0 b}$. By making the dominant pole of the closed loop system faster than the open loop one ($z_{cl} = \gamma a$), with $0.9 \leq \gamma \leq 1$ in order to avoid high oscillations), one obtains:

$$g_0 = \frac{1-(2(\gamma a - 0.5))^2}{4b} \tag{7.3}$$

which gives another relation of the controller parameters related to the system ones (remember that $g_1/g_0 = -a$).

With these relationships and taking into account the possible values achievable by a, b and γ, the range of possible values of controller parameters can be obtained:

$$\frac{g_1}{g_0} = -a = -\frac{T_I}{T+T_I}$$

Then $T_I = \frac{aT}{(1-a)}$, obtaining from the relationship $0.8 \leq a \leq 0.95$ the following inequalities: $156 \leq T_I \leq 741$ (s). With equation (7.3) and the relationship $g_1/g_0 = -a$, $K_{PI} = \frac{\gamma a^2}{b}(1-\gamma a)$ can be obtained, which gives $0.188 \leq K_{PI} \leq 3.65$.

As has been mentioned, nominal controller parameters are needed to implement an IFPIC. The conservative fixed PI controller used as a backup controller in the starting phase of the operation has been used as the nominal one[1]. This conservative controller gives the response shown in Figure 7.2 (dotted-line) and is included in the family of possible controllers, being designed with $a = 0.95$, $b = 0.05$ and $\gamma = 0.95$, which gives values of $g_0 = 1.65$ and $g_1 = -1.56$ ($K_{PI} = 1.56$ and $T_I = 675$ s).

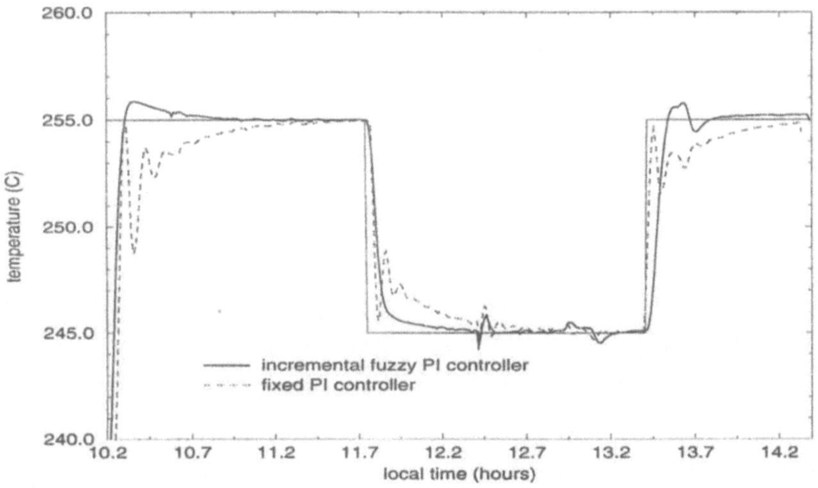

Fig. 7.2. Comparison between fixed PI control and incremental fuzzy PI control

In order to improve the behaviour of this controller in the normal operation, the fuzzy logic approach is used. If the mentioned controller is used as the nominal one, the controller parameters during the operation are given by:

$$\begin{aligned} K_{PI} &= K_{PI_{nominal}} + \Delta P\{E, \Delta E\}_P \\ T_I &= T_{I_{nominal}} + \Delta P\{E, \Delta E\}_I \end{aligned}$$

where values ΔP^{PI} are calculated by using a fuzzy matrix which is built taking into account maximum and minimum values of controller parameters (that have been slightly modified using the experience in the operation from the ones obtained in the theoretical development to avoid extremely low or fast responses: $0.5 \leq K_{PI} \leq 3.5$ and $150 \leq T_I \leq 675$ s).

The control scheme is shown in Figure 7.3. The error between the output of the plant and the set point signal (e) and its increment (Δe) are considered

[1] Obviously, the controller performance could be improved by choosing a better nominal controller. The main reason which justified the use of the conservative controller was to show the plant responsibles how the performance of a PI controller widely used at the plant could be easily improved by means of a fuzzy logic inference mechanism

to be the inputs for the fuzzy inference mechanism. The output variables of the fuzzy controller are the increments in the controller parameters.

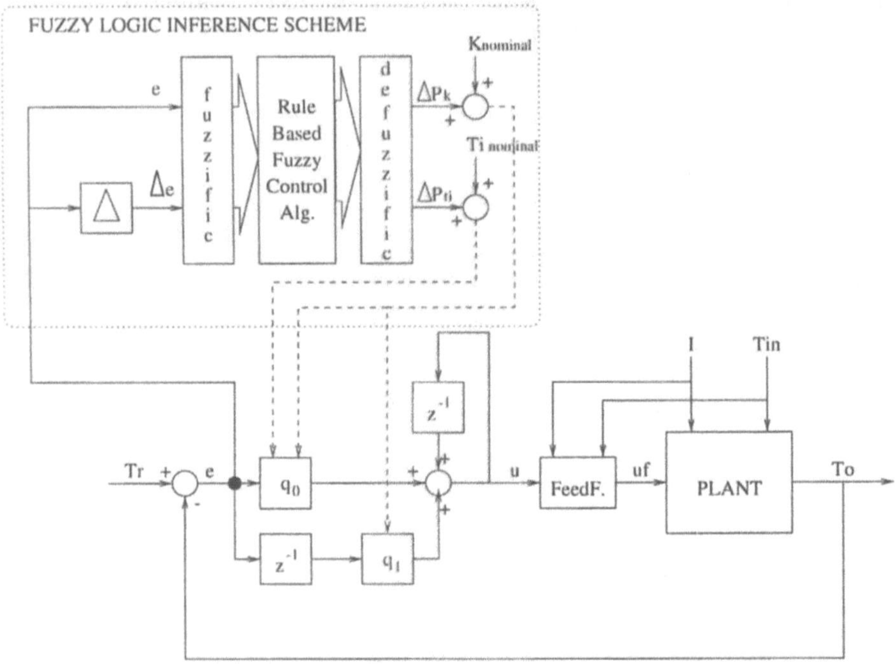

Fig. 7.3. Incremental Fuzzy PI Control scheme (IFPIC)

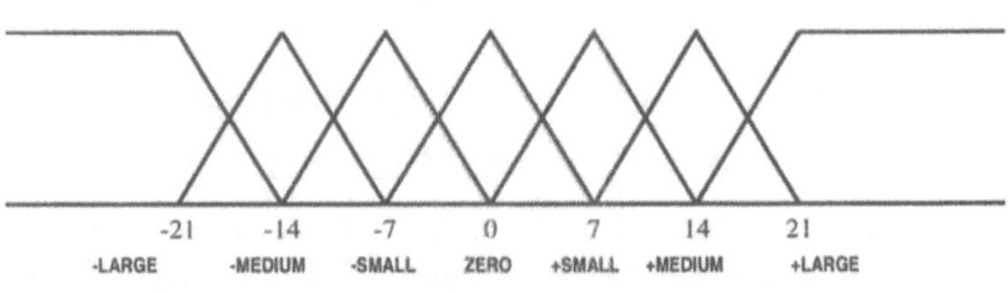

Fig. 7.4. Membership functions of the antecedents (e and Δe)

TP and TPE systems have been chosen for the inference mechanism. A triangular decomposition of the input universes of discourse can be seen in Figure 7.4. As can be seen, the extremes of the input universes of discourse (corresponding to e and Δe) are $[-21, 21]$. Those of the output have been chosen to be $[-1, 1]$ (corresponding to ΔK_{PI}) with centers $[1.95, 1.40, 1.20, 1.0, -0.1, -1.0]$ and $[-525, 0]$ (corresponding to ΔT_I), with centers $[0, -220, -250, -320, -450, -525]$. In this case (two input - single output system), the rules take the form (e.g. with K_{PI}):

IF *error* is **positive large** AND the *change in error* is **negative small** THEN make the K_{PI} *change* **positive large**.

where terms small, large, medium, etc are used to describe the fuzzy variables and the control action. As mentioned previously, because of the partial matching attribute of fuzzy control rules and the fact that the preconditions of rules do overlap, four rules can fire at the same time.

Assuming that the *algebraic product* operator is used as the conjunction operator, each rule recommends an action ($\Delta P_{i,j}^{PI}$, P^{PI} denoting a PI parameter) with a membership function $\mu_{\Delta P_{i,j}^{PI}} = \mu_{e_i} \cdot \mu_{\Delta e_j}$, where μ_{e_i} and $\mu_{\Delta e_j}$ are calculated by equation (7.1). The methodology used in deciding what control action should be taken results in the firing of four rules. The combination of these rules produces a nonfuzzy action ΔP^{PI}, which is calculated using the weighted averaging defuzzification method. Then, the resulting action can be calculated by,

$$\Delta P^{PI} = \frac{\sum_i^{i+1} \sum_j^{j+1} (\mu_{\Delta P_{i,j}^{PI}} \cdot \Delta P_{i,j}^{PI})}{\sum_i^{i+1} \sum_j^{j+1} \mu_{\Delta P_{i,j}^{PI}}} \tag{7.4}$$

and as in this case $\sum_i^{i+1} \sum_j^{j+1} \mu_{\Delta P_{i,j}^{PI}} = 1$, we have,

$$\Delta P^{PI} = \sum_i^{i+1} \sum_j^{j+1} (\mu_{\Delta P_{i,j}^{PI}} \cdot \Delta P_{i,j}^{PI})$$

Taking into account these maximum and minimum values and using the exposed inference mechanism (using triangular partitions in the input and output spaces and the weighted averaging defuzzification method), a description of the fuzzy control system can be obtained. A two-dimensional FAM for each one of the controller parameters can be used (Table 7.1).

Here we also show gain and integral time constant increment surfaces, obtained after a few simulations. The midpoints were slightly changed from the initial chosen values in order to achieve good results in the controller performance.

Results showing the behaviour obtained with this approach have been given in Figure 7.2, comparing them to the behaviour obtained with the fixed nominal PI controller. As can be seen, the overall performance is greatly improved, with faster rise times and with less than 3 degrees of overshoot when great changes in the set point signal occur.

7.2.2 Plant results

The proposed control algorithm has been tested at the solar power plant with very acceptable results. These results show the robustness inherent to the obtained nonlinear control in spite of set point changes. Figure 7.6 and 7.7 show the characteristic variables in a test in the presence of passing clouds. As can be seen, the outlet oil temperature follows the reference temperature

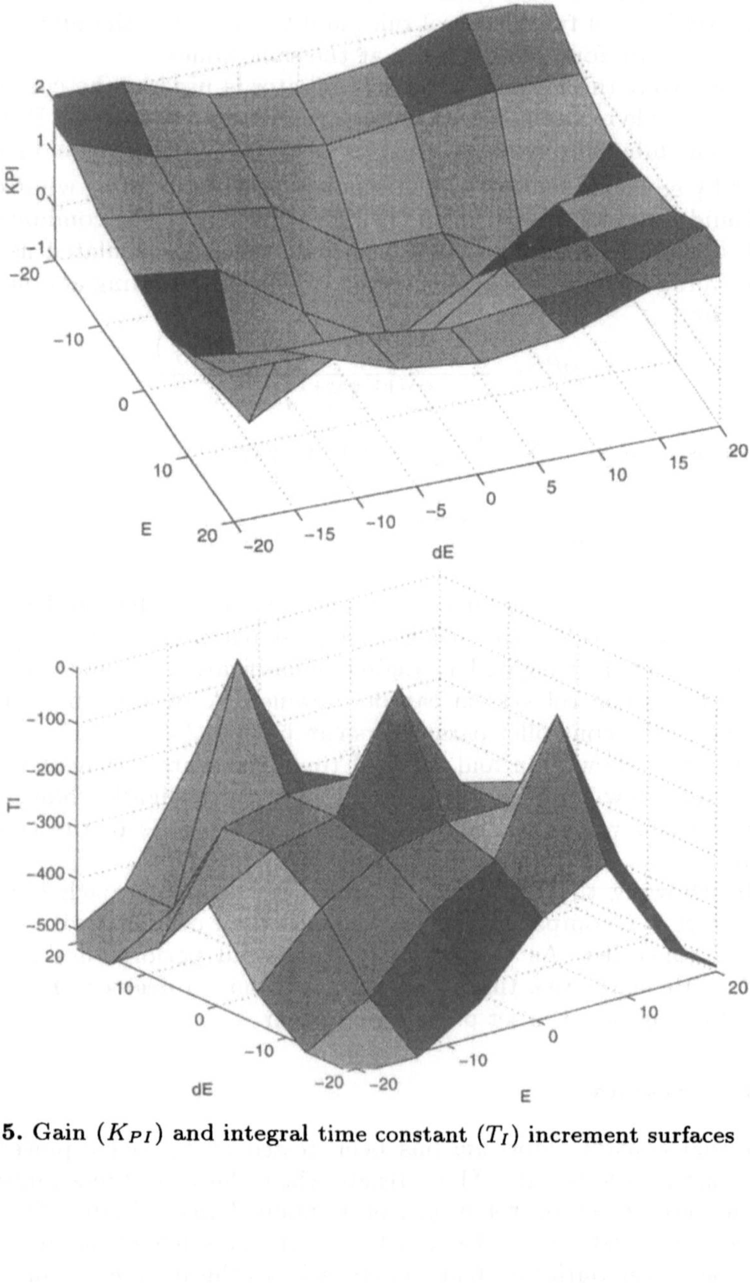

Fig. 7.5. Gain (K_{PI}) and integral time constant (T_I) increment surfaces

Table 7.1. FAM of the controller parameters

Kpi

Δe \ e	-L	-M	-S	Z	+S	+M	+L
-L	+EL	+EL	+L	+M	+S	+L	+EL
-M	+EL	+L	+M	+S	-S	+M	+EL
-S	+L	+M	+S	-ES	-ES	+M	+L
Z	+M	+S	-ES	-S	-ES	+S	+M
+S	+L	+M	-ES	-ES	+S	+M	+L
+M	+EL	+L	-S	+S	+M	+L	+EL
+L	+EL	+L	+S	+M	+L	+EL	+EL

Ti

Δe \ e	-L	-M	-S	Z	+S	+M	+L
-L	-EL	-EL	-L	-M	-S	-L	-EL
-M	-EL	-L	-M	-S	Z	-M	-EL
-S	-L	-M	-S	-ES	-ES	-M	-L
Z	-M	-S	-ES	Z	-ES	-S	-M
+S	-L	-S	-ES	-ES	-S	-M	-L
+M	-EL	-L	Z	-S	-M	-L	-EL
+L	-EL	-L	-S	-M	-L	-EL	-EL

EL: extra-large; L: large; M: medium; S: small; ES: extra-small; Z: zero

without great changes in the oil flow that could deteriorate this actuator. The estimated controller gain K_{PI} and integral time T_I evolution are shown in Figure 7.7, following the tendency pointed out by the rules used to design the IFPIC.

Figures 7.8 and 7.9 show the operation with the IFPIC on a day with strong disturbances caused by large passing clouds which produce drastic changes in the direct solar radiation level. Under these conditions, the effect of the feedforward controller in series with the IFPIC allows for very acceptable behaviour even under such extreme conditions (with deviations from the set point of less than 10 degrees with long time solar radiation changes greater than 300 W/m^2). Figure 7.9 shows the evolution of the direct solar radiation, oil flow and controller parameters.

Finally, the operation on a clear day (Figure 7.10) is presented, including the starting phase. The behaviour of the controlled system is very acceptable throughout the whole range of operating conditions, even in the starting phase (where the fuzzy inference mechanism operates at the extremes of the FAMs) with a thermal gradient greater than 150°C. Figure 7.11 shows the evolution of the direct solar radiation, oil flow and controller parameters.

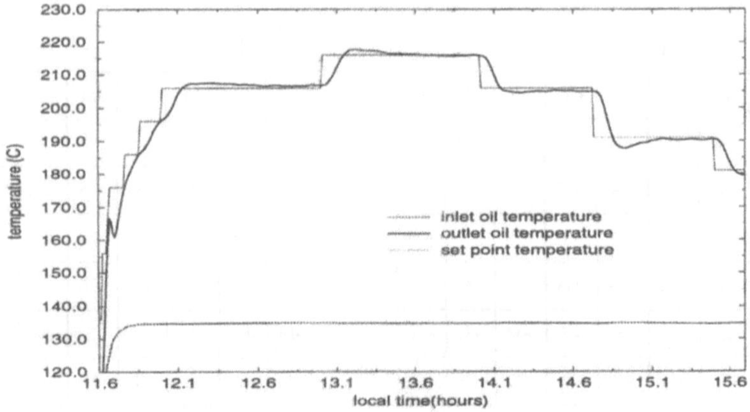

Set point, inlet and outlet oil temperatures (°C)

Fig. 7.6. Test with the IFPIC controller (17/01/95)

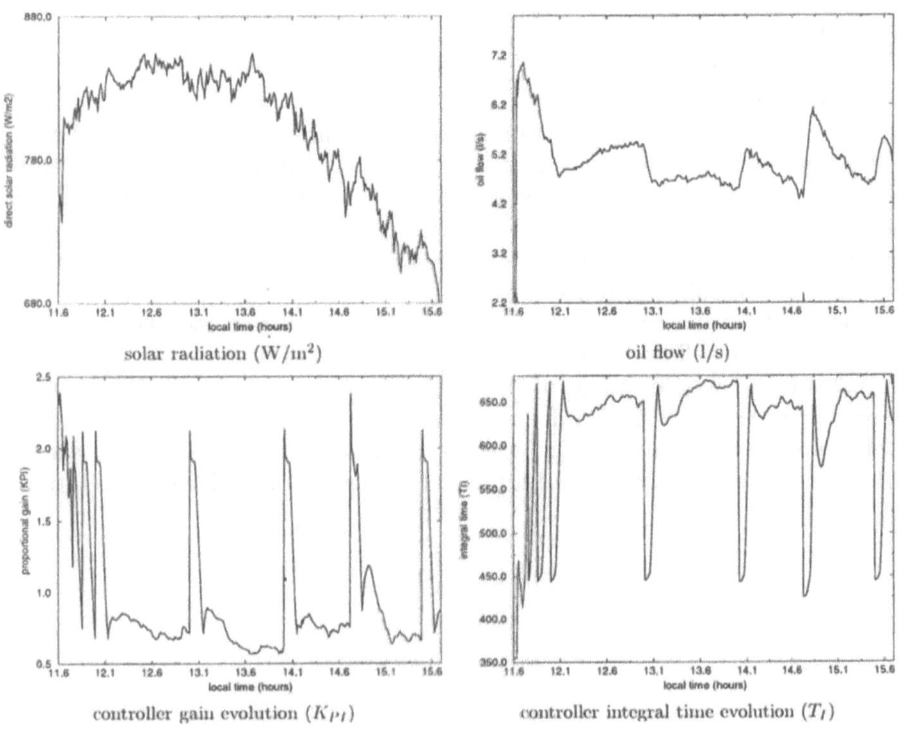

Fig. 7.7. Test with the IFPIC controller (17/01/95)

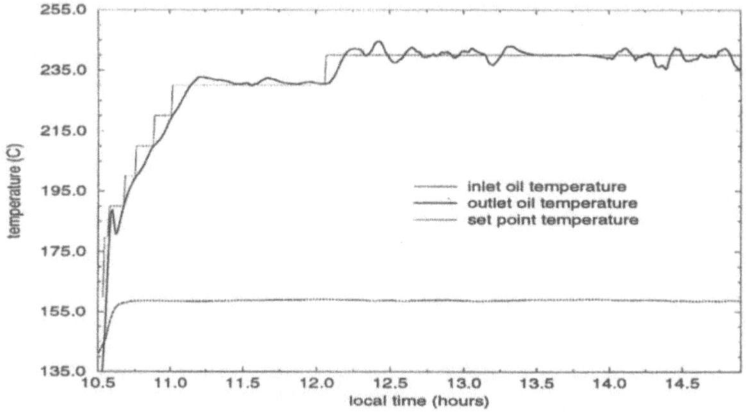

Set point, inlet and outlet oil temperatures (°C)

Fig. 7.8. Test with the IFPIC controller (03/03/95)

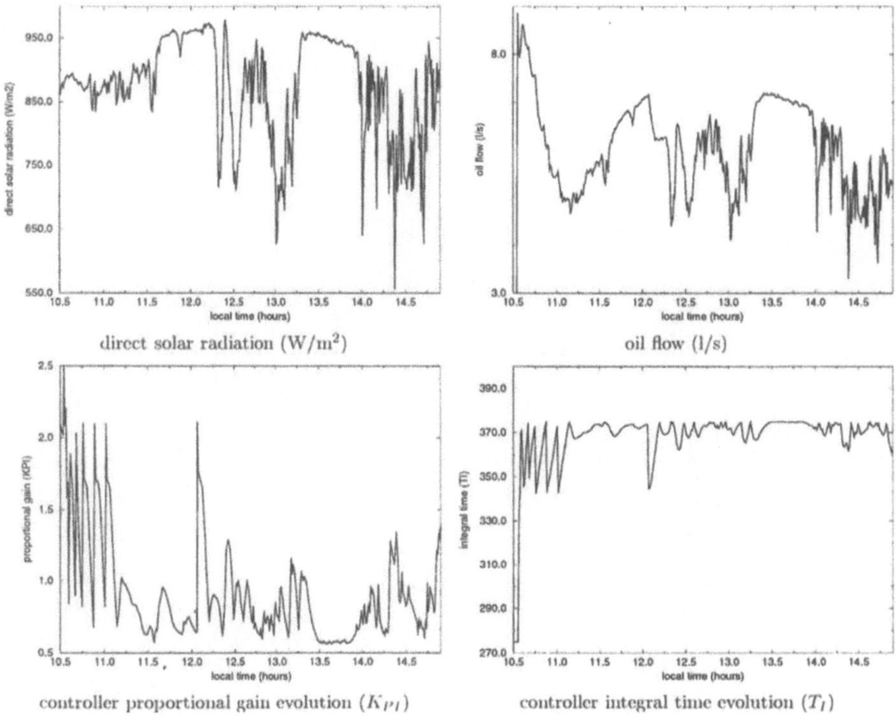

direct solar radiation (W/m^2)

oil flow (l/s)

controller proportional gain evolution (K_{PI})

controller integral time evolution (T_I)

Fig. 7.9. Test with the IFPIC controller (03/03/95)

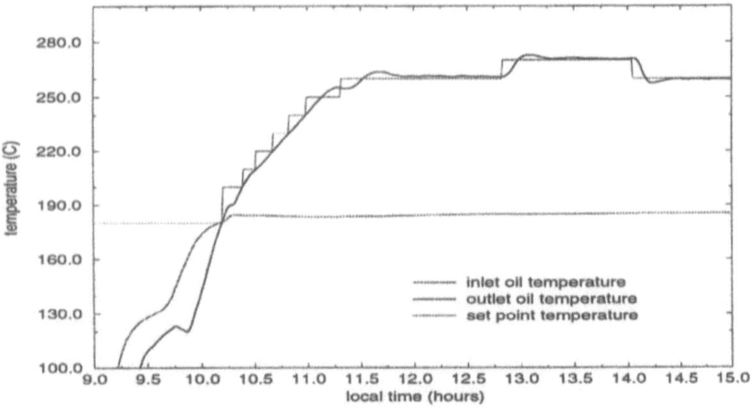

Set point, inlet and outlet oil temperatures (°C)

Fig. 7.10. Test with the IFPIC controller (04/03/95)

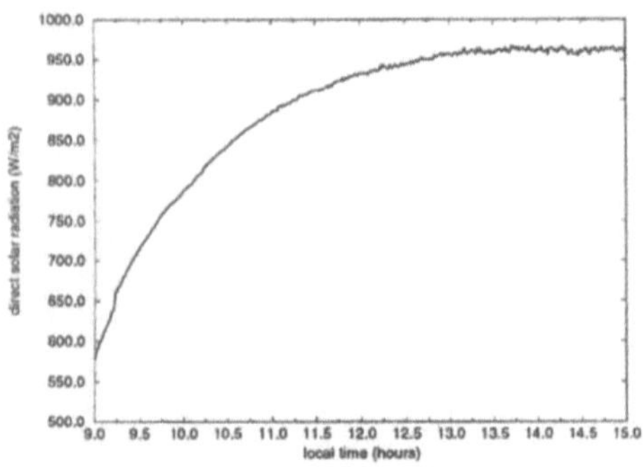

direct solar radiation (W/m²)

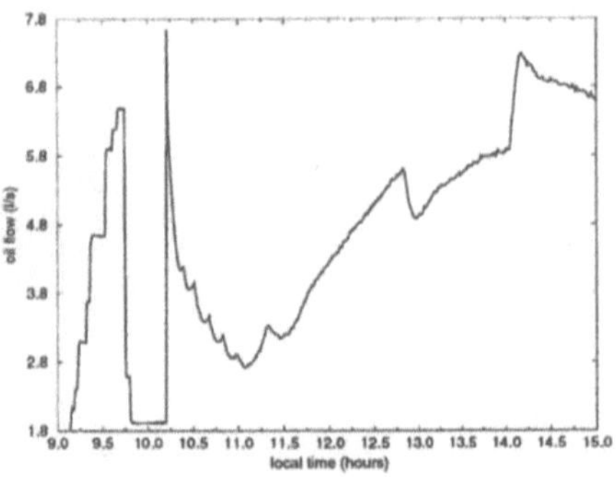

oil flow (l/s)

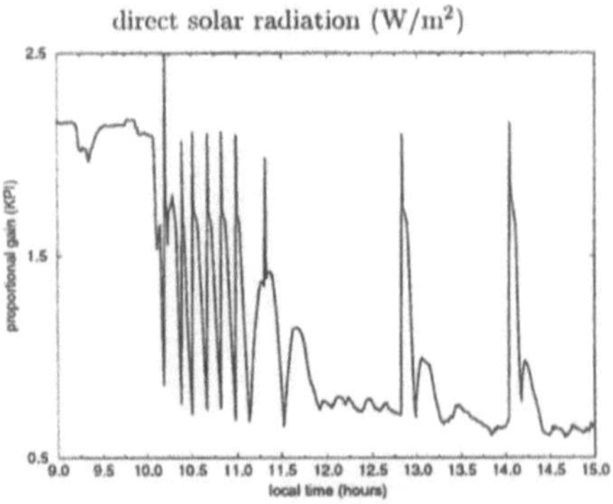

controller proportional gain evolution (K_{PI})

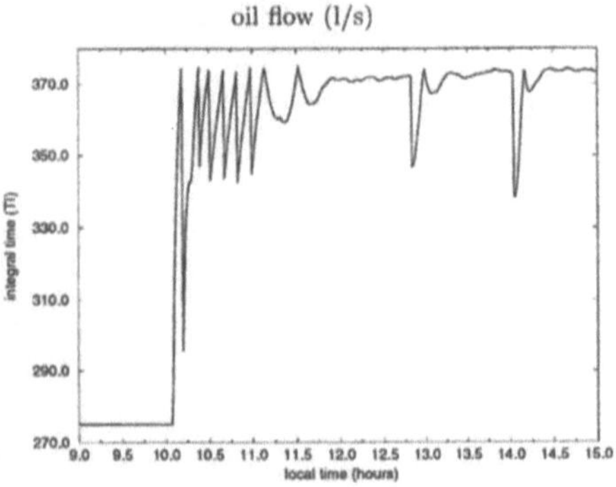

controller integral time evolution (T_I)

Fig. 7.11. Test with the IFPIC controller (04/03/95)

7.3 Fuzzy logic controller (FLC)

This section introduces a fuzzy logic controller which directly calculates the control signal. In this case, the FLC has been applied in an incremental form in series with the feedforward controller, that is, the signal obtained from the FLC is the increment needed in the control signal (reference temperature to the feedforward controller) to provide a desired behaviour (the output universe is related to control increments).

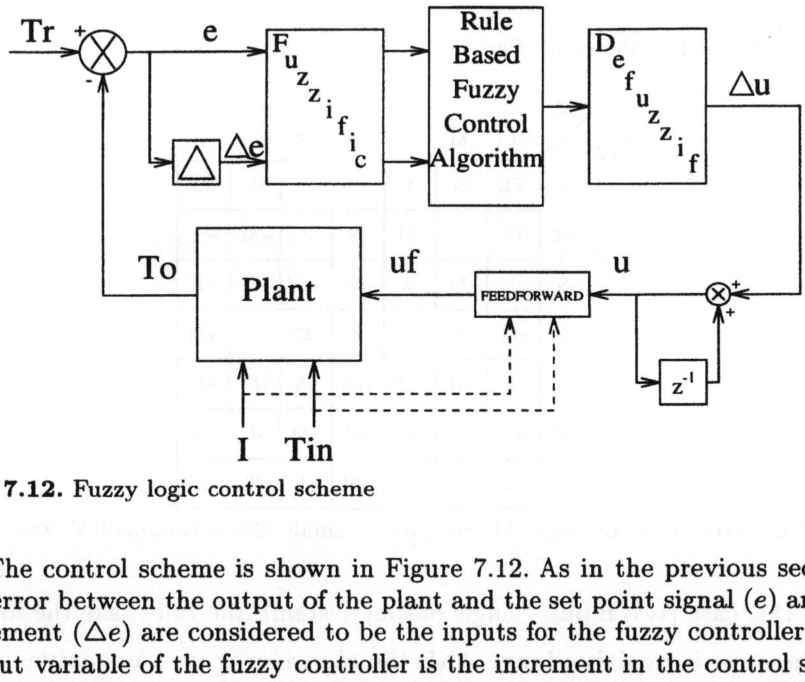

Fig. 7.12. Fuzzy logic control scheme

The control scheme is shown in Figure 7.12. As in the previous section, the error between the output of the plant and the set point signal (e) and its increment (Δe) are considered to be the inputs for the fuzzy controller. The output variable of the fuzzy controller is the increment in the control signal (reference temperature for the feedforward controller). The fuzzy controller could give the control signal directly, but it has been chosen in incremental form in order to introduce an integral effect in the control action and to reduce the fuzzy partition of the output domain. TP and TPE systems have also been chosen for the inference mechanism. Results obtained with other control strategies indicate that good results and some degree of robustness could be achieved by using an FLC with TP-TPE systems.

The triangular decomposition of the input universes of discourse has been shown in Figure 7.4. Again, the extremes of the input universes of discourse (corresponding to e and Δe) are $[-21, 21]$. Those of the output have been chosen to be $[-10, 10]$ (corresponding to Δu), with centers $[\pm 10, \pm 6, \pm 4, \pm 2, \pm 0.7, 0]$. The domains under consideration are not normalized. The discretization of the universes of discourse plays an important role in the final behaviour of the controlled system. In this case, the discretization has been selected taking into account the operating range of possible oil flows (between 2 and 12 l/s), the outlet temperature (with a maximum of 300 degrees) and previous knowledge of the plant dynamics. As is commented below, there is great experience in the control of the plant with other control approaches. This fact, and the existence of the nonlinear distributed parameter model for simulation purposes, allows for a good discretization of the universes of discourse and the testing of the controller behaviour before its real implementation in the plant. In this way, few changes were made when implementing the controller in the plant.

Table 7.2. FAM of the controller

Δe \ e	-L	-M	-S	Z	+S	+M	+L
-L	-EL	-EL	-L	-M	-S	+L	+EL
-M	-EL	-L	-M	-S	Z	+M	+EL
-S	-L	-M	-S	-ES	+ES	+M	+L
Z	-M	-S	-ES	Z	+ES	+S	+M
+S	-L	-M	-ES	+ES	+S	+M	+L
+M	-EL	-L	Z	+S	+M	+L	+EL
+L	-EL	-L	+S	+M	+L	+EL	+EL

EL: extra-large; L: large; M: medium; S: small; ES: extra-small; Z: zero

In this case (two input - single output system), the rules take the form:

IF the *error* is **positive large** AND the *change in error* is **negative small** THEN make the *control signal change* **positive large**.

Terms small, large, medium, etc are used to describe the fuzzy variables and the control action. As mentioned previously, due to the partial matching attribute of fuzzy control rules and the fact that the preconditions of rules do overlap, four rules can fire at the same time.

Again the *algebraic product* operator is also used as the conjunction operator, each rule recommends a control action ($\Delta u_{i,j}$) with a membership function $\mu_{\Delta u_{i,j}} = \mu_{e_i} \cdot \mu_{\Delta e_j}$, where μ_{e_i} and $\mu_{\Delta e_j}$ are calculated by equation (7.1). The methodology used in deciding what control action should be taken results in the firing of four rules. The combination of these rules produces a nonfuzzy control action Δu, which is calculated using the weighted averaging defuzzification method, as mentioned before. Then, the resulting control action can be calculated by,

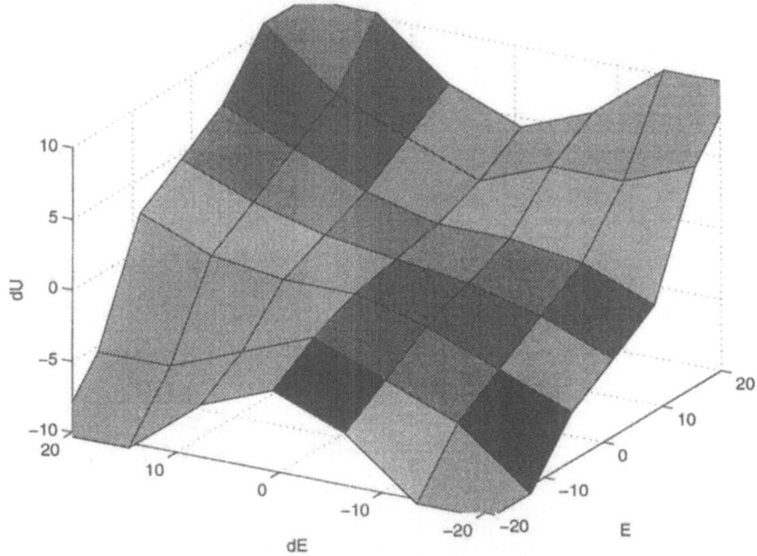

Fig. 7.13. Fuzzy control surface

$$\triangle u = \frac{\sum_i^{i+1} \sum_j^{j+1} (\mu_{\triangle u_{i,j}} \cdot \triangle u_{i,j})}{\sum_i^{i+1} \sum_j^{j+1} \mu_{\triangle u_{i,j}}} \tag{7.5}$$

and as in this case $\sum_i^{i+1} \sum_j^{j+1} \mu_{\triangle u_{i,j}} = 1$, we have,

$$\triangle u = \sum_i^{i+1} \sum_j^{j+1} (\mu_{\triangle u_{i,j}} \cdot \triangle u_{i,j})$$

A 2-dimensional FAM can also be used in this case (Table 7.2). The controller surface obtained from the resulting NI table can be seen in Figure 7.13.

7.3.1 FLC design procedure

The first step in the design procedure was to produce the FAM table shown in Table 7.2 and the centers of $\triangle u$ as $[\pm 20, \pm 15, \pm 10, \pm 6, \pm 3, 0]$. These values were chosen based on previous experience in controlling the plant. The resulting control was simulated with the nonlinear computer model and the results shown in Figure 7.14 (dotted line) were obtained. The simulations of the first FLC designed showed an undesirably high overshoot. Thus, the centers of the FAM were reduced, taking the following new values in the intermediate design: $[\pm 13, \pm 10, \pm 8, \pm 5, \pm 3, \pm 1.2, 0]$. With these values and the help of the nonlinear computer model, results shown in Figure 7.14 (dashed line) were obtained. This result can be considered to be a good one but,

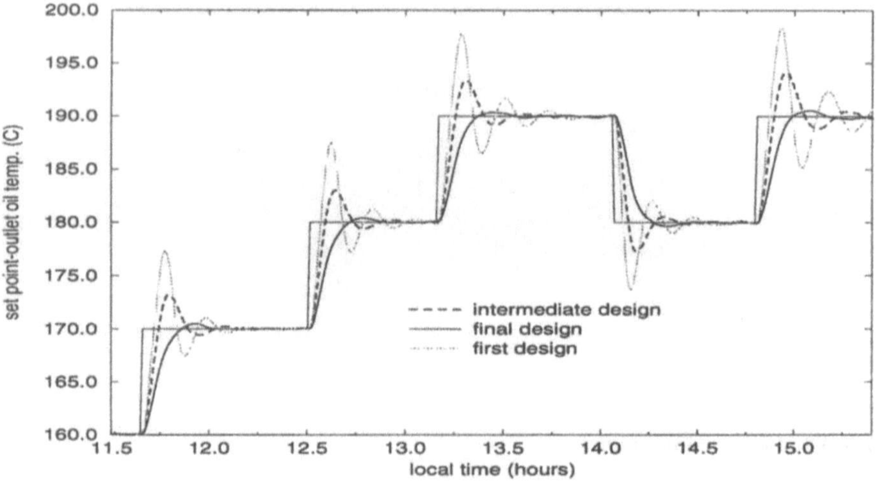

Set point and outlet oil temperatures (C)

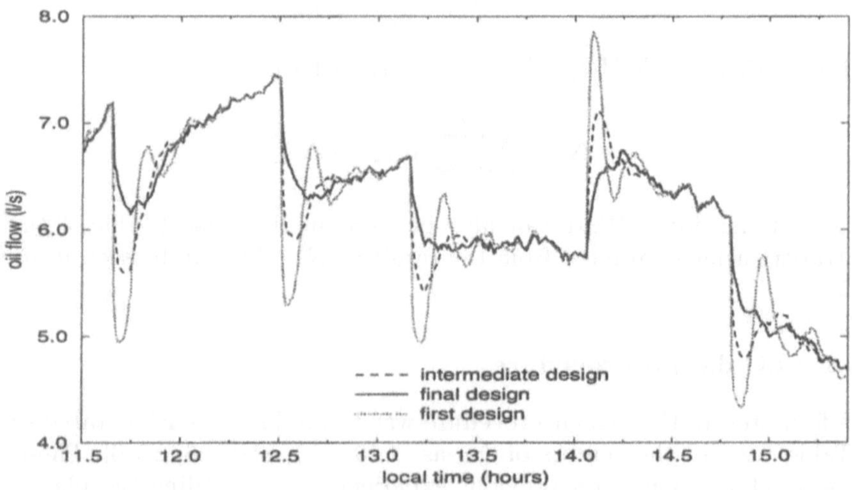

Oil flow (l/s)

Fig. 7.14. Simulation with different designs of the FLC

due to the characteristics of the solar plant, the main design objective is to achieve a response with low overshoot. In this way, the centers of the FAM were again diminished, obtaining the definitive ones mentioned above ($[\pm 10, \pm 6, \pm 4, \pm 2, \pm 0.7, 0]$). The simulation which corresponds to these values can also be seen in Figure 7.14 (solid line). This figure includes the values of the oil flow in order to show the control effort obtained with each one of the designs. These values have been used in the control of the distributed collector field, as shown in the next paragraph.

7.3.2 Plant results

The proposed control scheme has been tested at the plant. First of all, one aspect in the development of the FLC used here should be stressed, that is, the easiness of implementation and the very good results obtained in the first tests performed at the plant. This fact can be observed in Figure 7.15 and 7.16. Figure 7.15 corresponds to the step response test carried out on the 26th April, 1994 (outlet oil temperature, inlet oil temperature, set point temperature and solar radiation plots are provided). As can be seen, good set point tracking is obtained (even in cases of disturbances produced by small passing clouds at the end of the operation) with a rise time of less than 9 minutes in a step change of 10 degrees, and a similar response shape under all operating conditions.

Figure 7.16 (outlet oil temperature, inlet oil temperature, set point temperature, solar radiation and oil flow respectively) corresponds to another step response test (27/04/94), covering a wide range of oil flow conditions (from 5 to 7.5 l/s). As can be seen, very good results are obtained in all cases without oscillations in the system response. These were the first two tests performed at the actual plant. As can be seen, very good results were obtained with the initial design of the FLC. These curves show a small offset (less than 1.5 degrees) in the system response. The FLC has been developed using the error signal (e) and its increment (Δe), and the controller signal obtained is the control increment (which incorporates integral effect). In this way of implementation indicates that the system response should have null steady state error.

One possible explanation for the appearance of the offset in the output signal is the use of a wrong value of the mirror reflectivity (used in calculating the effective solar radiation) in the feedforward controller. Mirror reflectivity is measured once a week and so, if dust has accumulated on the mirror surfaces since the previous measurement, the real value of the mirror reflectivity can differ from the one last measured, used by the feedforward controller. This fact leads to an error in the outlet signal of the feedforward controller (oil flow demanded to the pump) that depends on the effective solar radiation. Theoretically, as the feedforward controller is placed in the control loop, this error should be corrected, but the integral term action is not fast enough to do so.

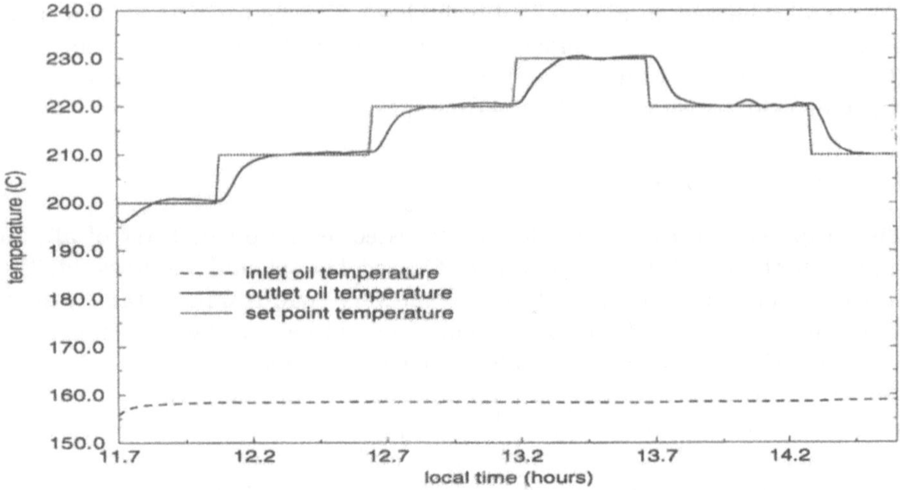

Set point, outlet and inlet oil temperatures (C)

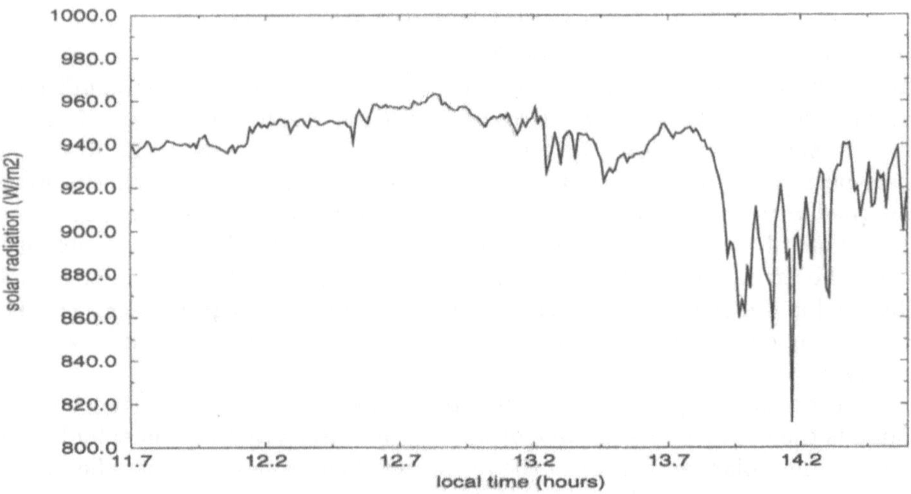

Solar radiation (W/m^2)

Fig. 7.15. Test carried out with the FLC controller (26/04/94)

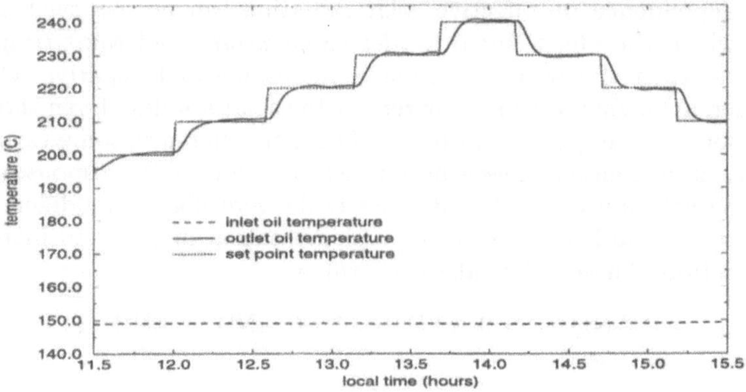

Set point, outlet and inlet oil temperatures (C)

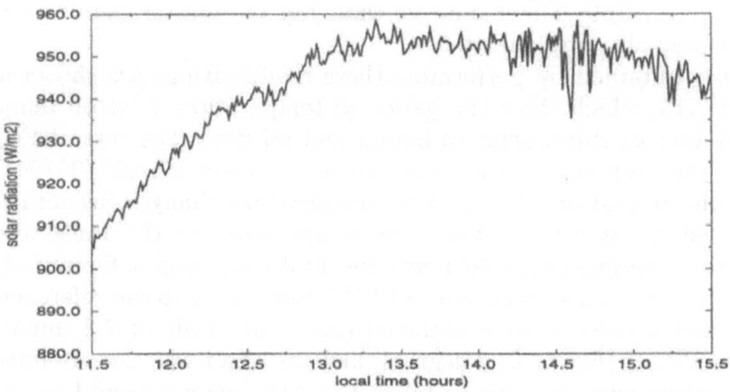

Solar radiation (W/m^2)

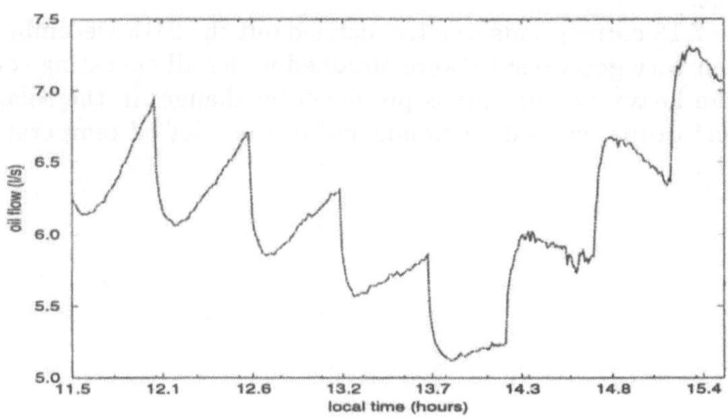

Oil flow (l/s)

Fig. 7.16. Test carried out with the FLC controller (27/04/94)

The dependence on effective solar radiation can be observed in Figure 7.16, in which the offset signal (*outlet temperature – set point temperature*) is positive when the solar radiation is increasing and negative when it is decreasing, achieving a value near zero at the solar midday. Even though this small offset is not important in terms of heat transference, some variations in the initial implementation were performed to achieve faster responses without offset. The values used in the inference table were slightly modified and two new rules were added in the FLC implementation in order to increase the integral action. These rules take the form:

IF $abs(e_k) > 2$ AND $e_k > 0$ AND $e_{k-1} > 0$ AND ... AND $e_{k-4} > 0$ THEN
$\Delta u_k = \Delta u_k + \phi$

where e_k is the actual error signal, e_{k-1} the previous one and so on. Δu_k is the actual control increment, *abs* indicates the absolute value and $\phi = \sum_{i=0}^{4} e_{k-i}/100$. The same rule changing the *greater than* sign for a *less than* sign was also implemented.

Results obtained by performing these modifications are shown in Figure 7.17 and 7.18, which show the outlet oil temperature, inlet oil temperature, set point temperature, solar radiation and oil flow. The test of Figure 7.17 provides the response during a whole day's operation (22/06/94). At the start of the operation, the inlet oil temperature changes because the oil is recirculated till it reaches the same temperature as the tank, and a step procedure is implemented to reach the first operating set point. Once the field reached $235°C$, step changes of $15°C$ were given to the reference. As can be seen, fast responses were obtained (rise time of about 7-8 minutes) with small overshoots (less than 2 degrees) and the offset was also eliminated. The control system was also able to reject the disturbances caused by changes in the solar radiation on the inlet oil temperature that was changed during the experiment.

Figure 7.18 corresponds to a test carried out the 21th December, 1994. As can be seen very good results were obtained under all operating conditions in spite of the heavy perturbations produced by changes in the solar radiation (200 W/m^2 drops) caused by clouds and in the inlet oil temperature (45 °C increment).

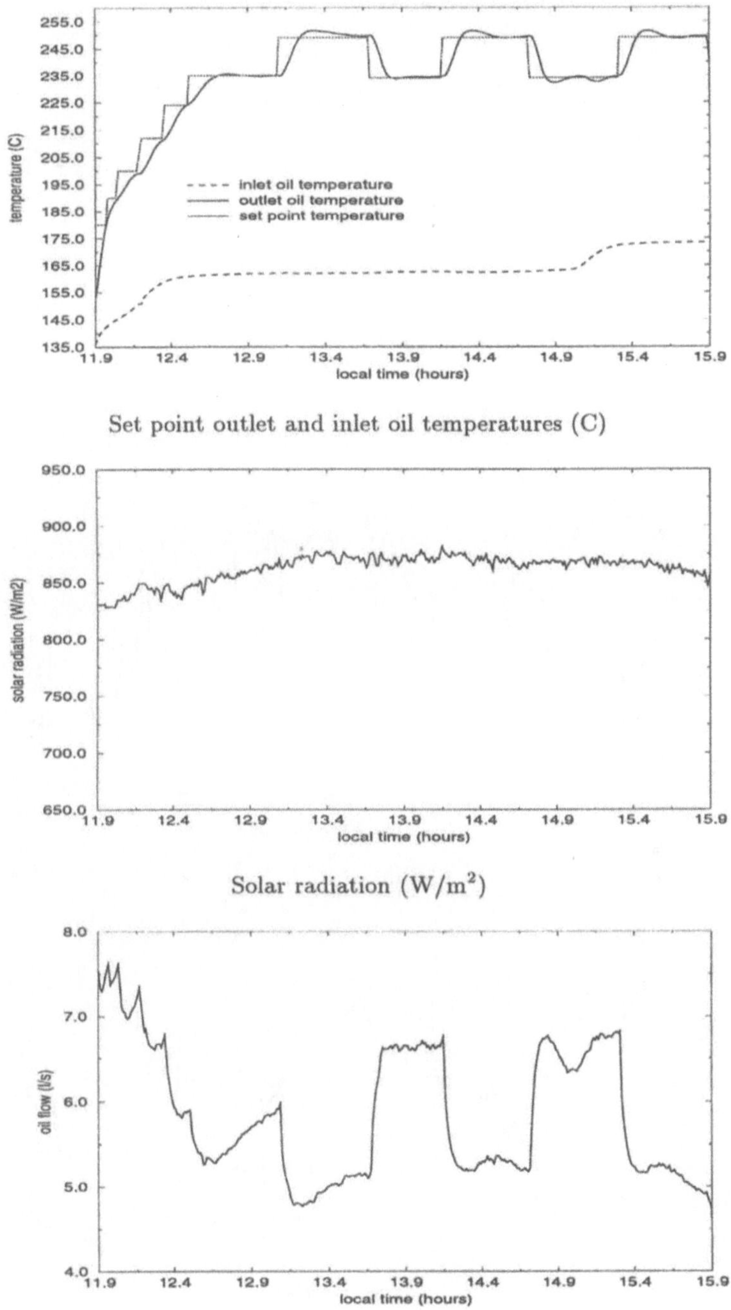

Set point outlet and inlet oil temperatures (C)

Solar radiation (W/m^2)

Oil flow (l/s)

Fig. 7.17. Test carried out with the FLC controller (22/06/94)

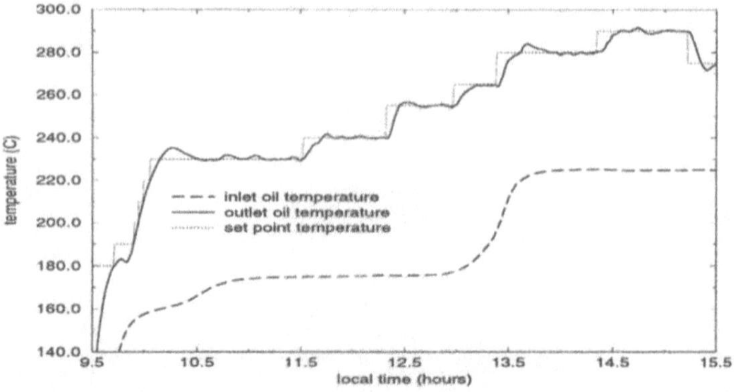

Set point outlet and inlet oil temperatures (C)

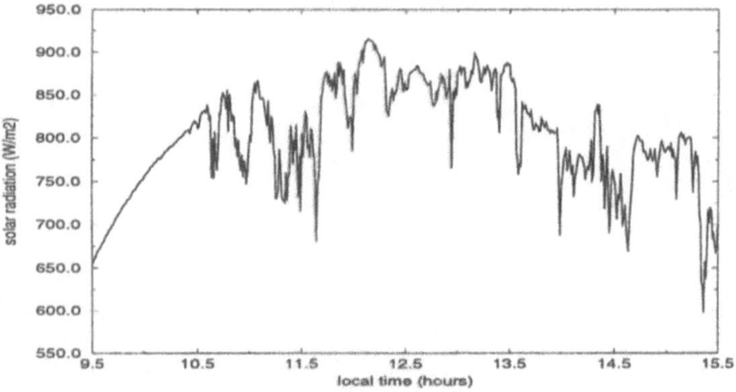

Solar radiation (W/m^2)

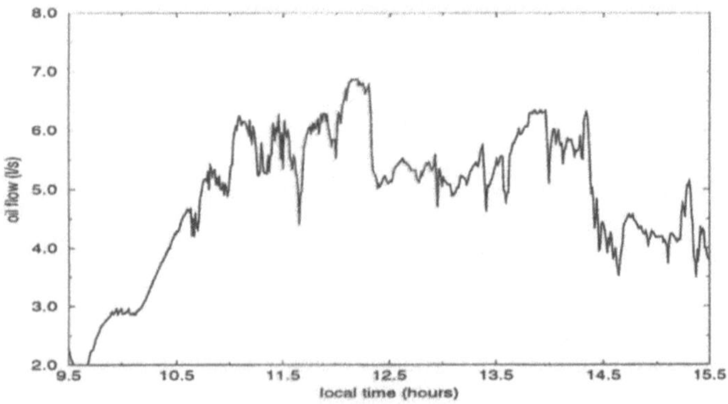

Oil flow (l/s)

Fig. 7.18. Test carried out with the FLC controller (21/12/94)

CHAPTER 8
SUMMARY AND CONCLUDING REMARKS

The aim of this chapter is to provide a brief summary of the main features of the different control schema included in the book. The most significant characteristics of control structures have been described in the chapters devoted to commenting on the development and implementation of the several controllers. Here, these characteristics are summarized and the performance and robustness properties of the controllers are compared. In order to carry out this comparison, the nonlinear distributed parameter simulation model has been used, as a way of obtaining more realistic results than those which would be obtained performing the study using different simplified linear models. The objective of the different advanced control schema when applied to the distributed collector field, is to provide acceptable behaviour throughout the entire range of operating conditions in spite of disturbances acting on the system. The simulation used here for performance and robustness analysis purposes covers different operating conditions and disturbances.

At this point, it is difficult to resist the temptation of making a comparative study between the different control strategies. From an objective viewpoint, it is impossible to perform a fair comparison between them, because they are based on different conceptual and methodological approaches. It is a common feature in many books and papers to perform a comparison between the *"splendid"* control strategy that the author is proposing and a fixed PID one. The *"unfortunate"* PID always looses the game. This is not the aim of the authors in this book. As has been seen in chapters 3 and 4, a fine-tuned PID has been obtained by modifying initial designs using accumulated experience[1] to provide a performance similar to that achieved with other advanced control strategies. Nevertheless, the performance of this controller deteriorates at low flow conditions. Adaptive, gain scheduling, nonlinear strategies, etc., which account for different operating points provide better results in these conditions than the PID one, but a comparison between them should not be too rigorous as the PID-based control strategy could be improved in the same

[1] As a matter of fact, it has been observed that the obtaining of a fine-tuned PID controller by different people working at the plant has been difficult, this being the main reason for why the plant has been operating using the PI backup controller shown in chapter 3 for such a long time.

way. In fact, results obtained with the adaptive PI controller and the GPC one (both based on low order linear models of the plant) are very similar. In these cases, the advantages of the implementation of the adaptive GPC controller arise when constraints are active, future references are known and more complex control models are used.

It is also important to point out that the different control strategies developed can be improved by changing tuning parameters (tuning knobs in GPC strategies, weighting matrices in LQG/LTR control, center values in fuzzy controllers, etc.) to provide better performance. In the various controllers, we have shown results obtained after the first few tests, because an important issue conditioning the use of advanced control strategies in the process industries is to assure a short commissioning time in the development and implementation of the controllers. One can always improve the performance of a control scheme, but it is important to find a trade–off between time devoted to designing the controller and the time in which an acceptable control scheme is implemented and fully operational at the plant. This fact is extremely important in those situations in which a reliable model of the plant does not exist and the tuning of the controllers has to be performed at the actual plant. Those responsible for the plant would not be very happy if they lost many days of operation or obtained poor results because the control expert was fine-tuning his splendid controller.

8.1 Performance indexes

Different simulation and plant results have been presented throughout the book. As indicated, the operating conditions cannot be reproduced exactly, as solar radiation, inlet oil temperature conditions and mirror reflectivity change from one day to the next. The simulator of the field has been used to perform a comparative study of the behaviour of each controller working under identical operating conditions [10]. In order to spare the reader the overwhelming results of these simulations, only a summary of the results is shown. The results obtained will be commented on and compared to those obtained at the plant. As has been mentioned, the objective of this simulation study is to confirm the results obtained at the actual plant covering the entire range of operating conditions and to perform a robustness analysis.

A reference file has been created containing the following variables: set point temperature, direct solar radiation, inlet oil temperature and ambient temperature. The simulations were performed at solar midday. The duration of the simulations was 750 sampling times (only the last 500 are shown in two indicative figures), equivalent to 8.125 operation hours. Obviously, if the simplification of considering a fixed solar hour was not taken into account, the feedforward controller would propose a saturated control signal during those hours in which the direct solar radiation was null or very low.

The test (Figures 8.1 and 8.2) consisted in two step changes (± 5 °C) in the outlet temperature set point followed by a sudden ± 100 W/m^2 perturbation in the solar radiation and a sudden step change (± 5 °C) in the inlet oil temperature.

For the set point temperature, three different tests were performed to allow the controllers to be tested under three different operating conditions (low, medium and high flow level), achievable with different values of the set point temperature.

Solar radiation and inlet oil temperature step changes are not so realistic, due to the fact that the variations in these variables present very different profiles during each test, depending on both meteorological and storage tank conditions. Throughout the book, simulations have been performed by using realistic operating conditions. In this case, the conditions are more ideal, and the solar radiation remains constant for a long time.

It is important to point out that all these simulations were carried out considering the real value of the mirror reflectivity ($R = 0.86$) to be that used by the feedforward controller or the nonlinear model (in the case of a nonlinear GPC controller). The robustness analysis which has also been carried out is based on the supposition of a badly estimated value of mirror reflectivity used by the feedforward terms.

8.1.1 Robustness analysis

The disturbances affecting the process can be classified into the two following classes:

- Fast or daily varying disturbances: those which can change during daily operation, such as solar radiation and inlet oil temperature.
- Slow varying disturbances: those factors which affect the energy collected by the field, but whose change is slow enough to consider them as constant during daily operation. An obvious example of this kind of disturbance is mirror reflectivity.

Both kinds of disturbances are taken into account by the feedforward controller and the nonlinear distributed parameter model of the plant. Inlet oil temperature and solar radiation variations are measured and updated on line, allowing the feedforward controller to compensate for these changes. The case of mirror reflectivity is different. This variable is measured once a week (or even every two weeks). Usually, this variable does not suffer great changes and so, the measured value can fortunately be used by the feedforward controller. Nevertheless, in adverse weather conditions prior to the daily operation (rain, dust carried by the wind, etc.), the real value of this variable can vary considerably from that measured a few days before.

Mirror reflectivity directly affects the solar energy collected and so, a wrong estimation of this variable can lead to systematic steady state errors. The physical variations of mirror reflectivity (R) usually lie within the range

$0.4 \leq R \leq 0.98$. Situations with poor reflectivity are quickly detected by the plant operators as the field is not able to appropriately heat the oil. Notice that a great difference between real and supposed mirror reflectivity can cause the feedforward controller to provide a control signal which will saturate the oil flow signal, impeding the control of the distributed solar collector field.

Due to these facts, control strategies should exhibit robust behaviour. A robustness analysis was carried out to study how mirror reflectivity uncertainty can affect the control schema presented in this book. Initially, the value of the reflectivity was changed to the limits over which the system becomes unstable. These values are indicated in the sections devoted to summarizing the main features of each controller. The nonlinear distributed parameter model of the field was used to simulate the plant behaviour considering reflectivity to vary from 0.1 to 1 (even covering values out of the previously mentioned range) and analyzing which value the control schema are unable to stabilize the output temperature of the field from. The same data used in the simulations performed with a nominal value of reflectivity ($R = 0.86$) were used to cover the entire range of possible oil flow conditions and other disturbances acting on the system.

As well as assuring stable performance, it can be of interest to analyze how the behaviour deteriorates with mirror reflectivity uncertainty. Notice that, as reflectivity multiplies the measured solar radiation, if the feedforward controller is fed with a smaller value of the reflectivity it will compute a flow smaller than the flow required and will force the feedback controller to correct these deviations. This will produce aggressive control actions and the relative robustness margins of the controlled system will be diminished, because operation with lower flow conditions than desirable causes the field to have a greater delay and the outlet temperature to achieve greater values than those indicated by the set point.

Figure 8.1 is indicative of the mentioned simulation tests carried out with all the controllers. In this case, the results obtained with the fixed PID controller are shown. As can be seen, performance deteriorates due both to operation under low or high flow conditions and to the uncertainty in mirror reflectivity. Comments about this and other simulations and results obtained at the plant are included in the following sections.

Prior to the analysis and use of the simulation results for summary purposes it is necessary to comment some general aspects:

- First of all, it is important to point out that, as was expected, in the simulations adaptive controllers generally produce better results than those achieved in the actual operation, due to the fact that the supposition of a main dependence between outlet oil temperature and oil flow is fulfilled when constant solar radiation and inlet oil temperature values are used.
- The nonlinear model-based GPC controller also shows better performance in the simulation than that obtained when operating at the solar plant, as

FIXED PARAMETERS PID CONTROLLER (MEDIUM FLOW CONDITIONS)

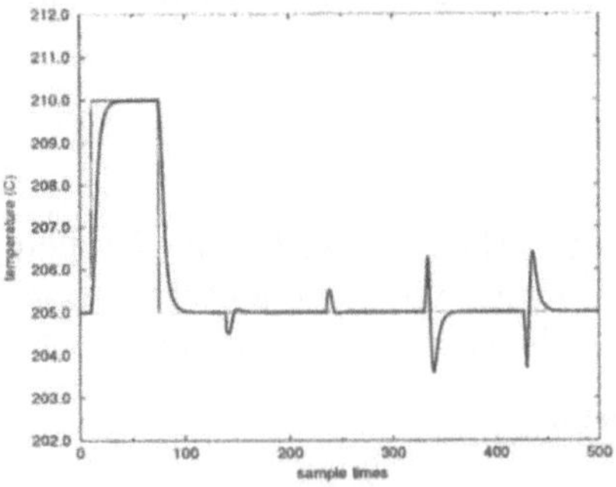

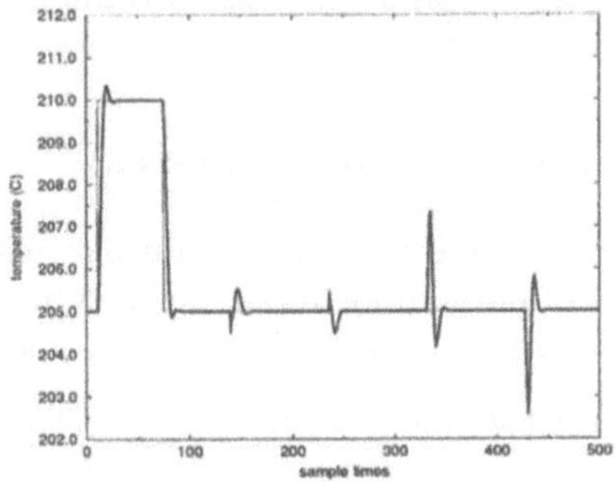

plant output with R=0.86 plant output with R=0.5

FIXED PARAMETERS PID CONTROLLER (LOW FLOW CONDITIONS)

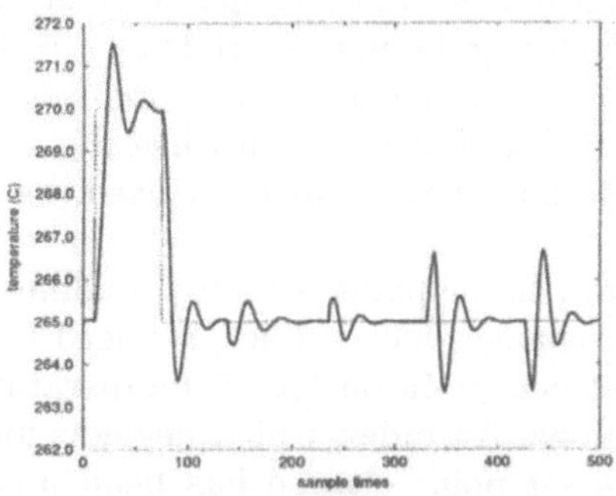

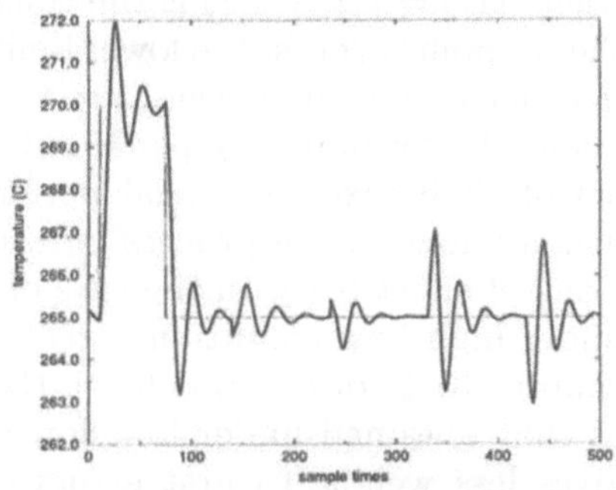

plant output with R=0.86 plant output with R=0.5

FIXED PARAMETERS PID CONTROLLER (HIGH FLOW CONDITIONS)

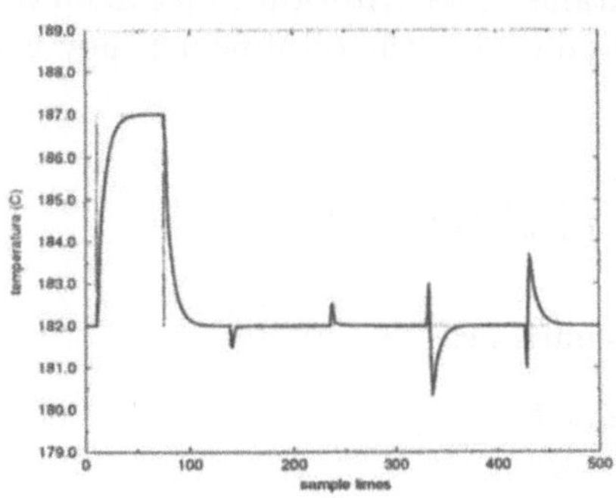

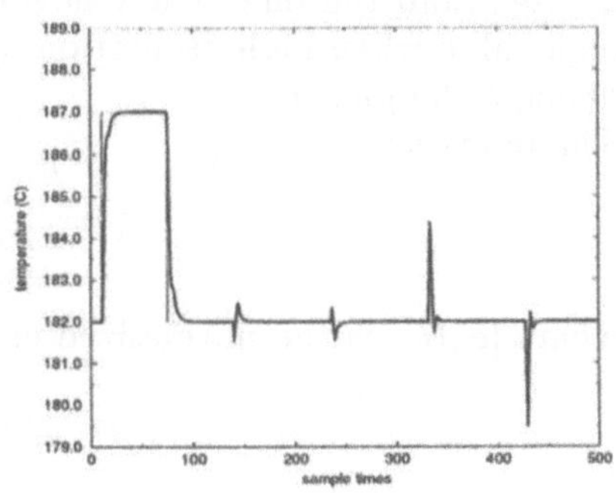

plant output with R=0.86 plant output with R=0.5

Fig. 8.1. Simulation with the fixed parameters PID controller the nonlinear model is quite similar to the simulator (only a few differences in the integration stage exist).

- For the other controllers, the behaviour obtained in several simulations is very similar to that obtained when operating at the distributed solar collector field.

8.1.2 Performance indexes

In order to analyze the main characteristics of the proposed control schema, a set of performance indexes has been used, which allows, to a certain extent, the controllers behaviour to be quantified. The chosen indexes have been:

- Integral time absolute error (ITAE criterion [40]):

$$\text{ITAE} = \sum_{t_{\text{change}}=0}^{t_{\text{new change}}} t\,|e(t)|$$

  where $e(t) = r(t) - y(t)$ is the tracking error at instant t. When a change in the set point occurs, the lower limit of the integral is made zero ($t_{\text{change}} = 0$), in order to give little weight to first errors. By using this index, oscillations about the set point are penalized. Although this criterion intrinsically covers other indexes, classical performance ones have also been used, as is usually done in the process industry.
- Control effort (CE): In the case of oil flow, as the plant is highly nonlinear, under high flow conditions greater flow changes have to be produced (see chapter 3) in order to achieve the same change in outlet oil temperature as that obtained under low flow conditions. An index which neglects and gives less weight to first errors after a set point change has been used, similar to that of the ITAE criterion, but also including error normalization to different flow conditions. If other indexes are used, a greater value of the criterion is always obtained at high flow levels as the flow step is also greater than the one at low flow conditions. The criterion must achieve a high value when high oscillations are produced in the oil flow, as they may damage the pump.
  The index is:

$$\text{CE} = \sum_{t_{\text{change}}=0}^{t_{\text{new change}}} t\,|e_u(t)|dt$$

  where $|e_u(t)|$ is the normalized error defined as:

$$e_u(t) = \frac{\Delta u(t)}{(u_{rp_i} - u_{rp_{(i-1)}})}$$

  and:
  $\Delta u(t) = u(t) - u(t-1)$: increment of oil flow demanded to the pump.

u_{rp_i}: required steady state oil flow.
$u_{rp_{(i-1)}}$: required steady state oil flow before the operating point was changed.

As can be seen, both indexes penalize the oscillations about a set point, both from the outlet oil temperature viewpoint and the oil flow demanded to the pump one. It is better to have smooth transients, even with small quasi-steady state errors (that is, the outlet oil temperature slowly approaches the set point, as in the case of the fixed PI backup controller long in use at the plant) than highly underdamped responses with large control efforts.

Other classical indexes are referred to in the text as a result of observations or measures in the time response of the plant. These classical indexes are:

- Overshoot (°C): When set point changes are performed, this index represents the maximum (minimum) value in degrees (°C) reached by the output over (under) the reference. If changes in solar radiation or inlet oil temperature occur, this index represents the maximum deviation (in absolute value) from the reference due to the disturbances.
- Rise time (min.): After a set point change, this index takes into account the time elapsed in minutes until the output of the system reaches 90% of the corresponding steady state value (this index includes dead times after the set point change, which is considered to be as the initial instant of the mentioned time lapse).
- Settling time (min.): The time elapsed in minutes between a set point change and the moment in which the output of the system lies inside a band of ±0.5°C around the new set point. If the steady state error is greater than this band, the settling time is measured in respect to the steady state value of the system output.
- Step response steady state error (°C): Difference in degrees (°C) between the reference and the output of the system in steady state.

8.2 Fixed PID controller

The set point tracking characteristics and disturbance rejection capabilities of this controller are excellent about the design point, but deteriorate when operating under other flow conditions (as can be seen in chapter 3 and Figure 8.1), mainly when low flow values are demanded. Analyzing the results obtained when operating at the actual plant it can be seen that this controller can produce an overshoot after a set point change of 10°C between 0 and 8 °C depending on the operating point, with rise times between 6.5 and 10 minutes and settling times between 6.5 and 25 minutes depending on the oil flow conditions. The CPU time consumed by this controller is quite small and does not constitute a drawback. The commissioning time of this controller is also small if acceptable behaviour is to be obtained. If a fine-tuned PID is required, a number of tests have to be performed to tune it, and this would be very time consuming if a precise nonlinear simulation model of the plant is not available.

From observations at the plant and results obtained in simulation, it can be seen that the controlled system tends to possess an oscillatory response under low flow conditions, because the residence time increases and the effect of the antiresonance modes easily affects the output temperature profile. Also, as flow decreases the delay of the system increases and the relative robustness margins decrease being the system more sensitive to pumped flow variations.

From the viewpoint of disturbance rejection capabilities, it can be seen that a great amount of the contribution needed to compensate for disturbances is provided by the series feedforward controller. The rest of the contribution is provided by the feedback controllers (most of all when coping with large low-frequency-gain controllers).

These results have been corroborated by performing the simulation testbed, obtaining similar values of overshoot, rise and settling time (a few smaller values are obtained due to the fact that the nonlinear distributed parameter model dynamics are slightly faster than those of the plant). Disturbance rejection is also acceptable.

Performance indexes have also been obtained and included in Table 8.1. Some remarks that can be pointed out are:

- Performance indexes achieve the greatest values at low flow conditions (performance deteriorates).
- Values of the ITAE criterion under high flow conditions are greater than those under medium flow conditions in the nominal case ($R = 0.86$), as the system evolution is slower.
- It was expected that the ITAE criterion with a wrong value of mirror reflectivity should be greater than those obtained with $R = 0.86$. This is not the case, due to the fact that the controller action is quite aggressive. In those cases in which a conservative response is obtained with $R = 0.86$ (medium and high flow conditions), the value of the ITAE criterion with $R = 0.5$ is less than that obtained with $R = 0.86$, as the control action makes the outlet oil temperature reach the set point faster. After analyzing these results, one would think that the initial controller was detuned. This is right because the PID controller was tuned to achieve a stable and acceptable result under a wide range of operating conditions. As can be seen in Table 8.1, values of the ITAE criterion under low flow conditions with $R = 0.5$ are quite high and the response (Figure 8.1) is stable but unacceptable.
- From the CE criterion viewpoint, its values are high under low flow conditions, as the chosen criterion penalizes oscillations of the control signal, that mainly occur under such conditions (also, high values of the ITAE criterion are obtained). In other conditions, the CE criterion gives adequate values which reflect smooth operation.

- Notice that the CE criterion is greater with a value of $R = 0.5$ than with $R = 0.86$ as the control signal is more vigorous, due to the considerations made above.

All these considerations are in general valid for the other control schema.

From the robustness viewpoint, the values of the estimated reflectivity R used by the feedforward controller which assure stable operation (R_{min} and R_{max}) have been obtained and are also included in Table 8.1.

Table 8.1. Fine-tuned PID controller: Reflectivity range and performance indexes

Reflectivity	**medium flow**	**low flow**	**high flow**
R_{min}	≥ 0.3	≥ 0.3	≥ 0.2
R_{max}	≤ 1.0	≤ 1.0	≤ 1.0
Performance indexes	$(\times 10^3)$	$(\times 10^3)$	$(\times 10^3)$
ITAE ($R = 0.86$)	24.9	93.5	31.7
ITAE ($R = 0.5$)	13.5	153.3	9.1
CE ($R = 0.86$)	1.2	3.3	1.3
CE ($R = 0.5$)	1.8	9.5	1.4

8.3 Adaptive GPC controller

The proposed control scheme should guarantee null steady state error after a step input because a CARIMA model has been used at the design stage. From the viewpoint of disturbance rejection capabilities, although the scheme is implemented using the series feedforward controller, the identification mechanism is sensitive to the injection inside the loop of dynamics not accounted for by the simplified linear model used in this control strategy, and the behaviour of the controlled system can deteriorate under conditions of strong disturbances. The commissioning time of the controller is quite short, as obtaining the simple adaptation equations is straightforward. The computational effort is also negligible as only a few operations have to be carried out.

When operating at the actual plant, the following performance characteristics have been observed: the overshoot after a set point change of 10°C is between 0 and 3 ^{o}C when operating with a weighting factor $\lambda \geq 6$ and depending on the operating point, with rise times between 5 and 10 minutes and settling times between 5 and 15 minutes. Acceptable disturbance rejection capabilities have also been observed, with the problems already mentioned which justify the inclusion of supervisory mechanisms in order to increase the robustness and stability of the identifier.

When performing the same simulation carried out with the fixed PID controller of the previous section, very similar (slightly better) tracking and disturbance rejection characteristics have been obtained when compared to actual plant results. The performance around medium flows is quite similar to that obtained with the PID controller, but the adaptive GPC controller achieves better performance when operating under high and low flow conditions, as is reflected in Table 8.2, which contains the performance criteria. The same general comments made in the previous case (fixed PID controller) apply here. Nevertheless, performance deterioration occurs when operating at low flow conditions and with a reflectivity value different to the nominal case, but this deterioration is much lesser (see Table 8.2) than that obtained with fixed controllers, as is to be expected from the use of an adaptation mechanism which can cope with different operating conditions.

At high flow conditions, the behaviour does not deteriorate, due to the ability of the controller to adapt itself to different operating conditions.

When operating around low flow values, oscillatory behaviour appears due to the factors commented (increase of delay and effect of antiresonance modes), as the model used (type A model) does not account for them, but these oscillations are smaller than those of the fixed PID controllers because the control system is operating with a low order model designed to cope with low flow conditions (this can be seen in the value of the ITAE criterion).

From the robustness viewpoint, in the case low order model based adaptive controllers, it is to be expected that the adaptation mechanism should cope with steady state gain variations (produced by uncertainty in reflectivity), as the plant being identified is constituted by the feedforward controller in series with the field. In this case, the performance deterioration is mainly due to changes in the delay of the field caused by operation under low flow conditions.

The values of the estimated reflectivity used by the feedforward controller which assure stable operation have been obtained from the robustness study, and are included in Table 8.2.

Table 8.2. Adaptive GPC controller: Reflectivity range and performance indexes

Reflectivity	**medium flow**	**low flow**	**high flow**
R_{min}	≥ 0.2	≥ 0.2	≥ 0.2
R_{max}	≤ 1.0	≤ 1.0	≤ 1.0
Performance indexes	$(\times 10^3)$	$(\times 10^3)$	$(\times 10^3)$
ITAE $(R = 0.86)$	25.1	54.5	31.7
ITAE $(R = 0.5)$	13.6	70.9	12.4
CE $(R = 0.86)$	1.5	5.3	1.3
CE $(R = 0.5)$	3.1	13.3	1.9

8.4 Robust adaptive GPC controller

From the theoretical viewpoint, the same considerations explained in the previous section for the case of the adaptive GPC arise in this case, except in the case of parameter identification, because due to the use of a robust identification mechanism the complete control scheme is less sensitive to the injection of disturbances or unmodelled dynamics. The controller has been developed by using a CARMA model of the plant and so, small steady state errors can appear in the plant response (if a wrong estimation is performed or if the chosen grid in the optimization algorithm is not small enough).

When this control scheme was implemented at the plant, overshoots between 2 and 3 °C were obtained, with rise times between 6.5 and 9 minutes and settling times between 6.5 and 12 minutes (sometimes due to slow steady state error compensation). This control algorithm shows good disturbance rejection capabilities. The test-bed simulation performed has confirmed plant results, even at those operating points not covered by several tests.

The main drawbacks of this controller are the conceptual difficulties inherent to the algorithm and the computational effort, which is the greatest of the control schema presented, needing 7 minutes to calculate the control signal (in a 486DX2 computer) if the algorithm is fully implemented (estimating all the parameters with control and prediction horizons of about 7 sampling times). The implementation shown in the book only consumes 1 second as parameter a is not estimated.

As has been mentioned before, when performing the simulations to study the main characteristics and robustness of the scheme, few differences were found between the standard and the robust GPC controllers. In real operation advantages of the robust design arise because the inclusion of a robust identification mechanism and the use of a MIN-MAX type of optimization algorithm provide a controlled system less sensitive to parameter identification errors of the chosen models.

The large computational effort required to perform the optimization leads to the selection of small control and prediction horizons, which usually produce a slower responses when compared to standard GPC algorithms. This can be slightly corrected by the modification of the weighting factor λ, but a trade–off must be found as is usually done when setting these tuning knobs.

Steady state errors can be produced with this type of controller because of its formulation using a CARMA model of the plant, mainly when great differences exist between the real reflectivity of the mirrors and that used in the feedforward controller, as has been corroborated by the robustness analysis. The ITAE criterion has been applied taking into account the steady state error of the plant (the error is considered to be null if the output is about the set point in steady state) to avoid this criterion achieve values which would not correspond to the controller features.

The disturbance rejection capabilities are similar to those of the standard GPC adaptive controller in the simulations performed. When operating at the solar plant, due to the persistence of the disturbances mainly due to solar radiation changes, the behaviour of the robust adaptive GPC improves that of the adaptive GPC, as the variability range of the estimated parameters decreases throughout time.

The values of the main performance criteria and the estimated reflectivity used by the feedforward controller which assure stable operation have been obtained and are shown in Table 8.3. In this case, it is important to point out that the robust nature of the controller produces smaller changes of the performance criteria when operating conditions are changed.

Table 8.3. Robust adaptive GPC controller: Reflectivity range and performance indexes

Reflectivity	medium flow	low flow	high flow
R_{min}	≥ 0.3	≥ 0.3	≥ 0.3
R_{max}	≤ 1.0	≤ 1.0	≤ 1.0
Performance indexes	$(\times 10^3)$	$(\times 10^3)$	$(\times 10^3)$
ITAE $(R = 0.86)$	27.2	53.1	29.5
ITAE $(R = 0.5)$	16.2	72.5	14.6
CE $(R = 0.86)$	1.4	5.4	1.3
CE $(R = 0.5)$	3.0	5.5	2.6

8.5 Gain scheduling GPC controller

This control strategy uses linear plant models which account for the presence of antiresonance modes and, moreover, it allows for a fast modification of the controller parameters to cope with changes in process dynamics, and therefore a fast and well damped set point tracking without steady state errors is obtained. As the model used accounts for the antiresonance characteristics, these types of oscillations are not present.

The good disturbance rejection capabilities of this control strategy are to be mentioned, obtained as a combination of the action of the series feedforward controller and the fast accommodation of the controller parameter values to new dynamic situations.

The observed overshoots after a set point change of 10°C are between 0 and 1.5 °C depending on the operating point, with rise and settling times of between 6 and 11 minutes. The computational effort is negligible as only few operations have to be performed. The commissioning time of this controller is also small. The main drawback is the necessity to perform PRBS tests at the plant that can be performed in one or two days, depending on the operating points to be characterized. The achievement of a robust and stable performance is not assured and so simulations or plant tests have to be performed, covering a full range of operating conditions, before the final implementation for daily operation.

As has been observed in the plant results, this controller shows excellent performance in spite of set point changes and disturbances acting on the system throughout the whole range of operating conditions, without oscillations due to the excitation of the antiresonance modes or changes in plant dynamics with the operating conditions, because these have been taken into account at the design stage of the algorithm.

Simulation results are similar to those obtained in the actual operation, as was to be expected. As this controller provides excellent results, a simulation is provided to show its capabilities for coping with different operating conditions (Figure 8.2).

From the robustness viewpoint, the gain scheduling controller again shows excellent behaviour in spite of gain uncertainties, as it has been designed using different models of the plant made up of the feedforward controller in series with the field (as is done with GPC adaptive controllers).

The values of the estimated reflectivity used by the feedforward controller which assure stable operation are included in Table 8.4.

Table 8.4. Gain scheduling GPC controller: Reflectivity range and performance indexes

Reflectivity	**medium flow**	**low flow**	**high flow**
R_{min}	≥ 0.1	≥ 0.1	≥ 0.1
R_{max}	≤ 1.0	≤ 1.0	≤ 1.0
Performance indexes	$(\times 10^3)$	$(\times 10^3)$	$(\times 10^3)$
ITAE $(R = 0.86)$	21.6	43.0	20.4
ITAE $(R = 0.5)$	10.5	30.3	5.8
CE $(R = 0.86)$	0.7	2.3	0.7
CE $(R = 0.5)$	1.2	4.2	1.4

8.6 Nonlinear GPC controller

Due to the inclusion of a nonlinear model of the plant valid throughout the entire range of operating conditions, a fast set point tracking is obtained with null step response steady state error if a CARIMA model is used to calculate the *forced response* of the system and an incremental direct solar radiation prediction model is implemented. This scheme allows for compensating disturbances acting on the system, as the nonlinear model takes into account their corresponding dynamics.

When implemented at the plant, measured overshoot after a set point change of 10°C is between 0 and 3 °C, with rise times between 5 and 6 minutes and settling times between 5 and 13 minutes. Disturbance rejection capabilities are good. The commissioning time of this controller is small if a reliable nonlinear plant model is previously provided. If this is not the case,

GAIN SCHEDULING GPC CONTROLLER. MEDIUM FLOW CONDITIONS

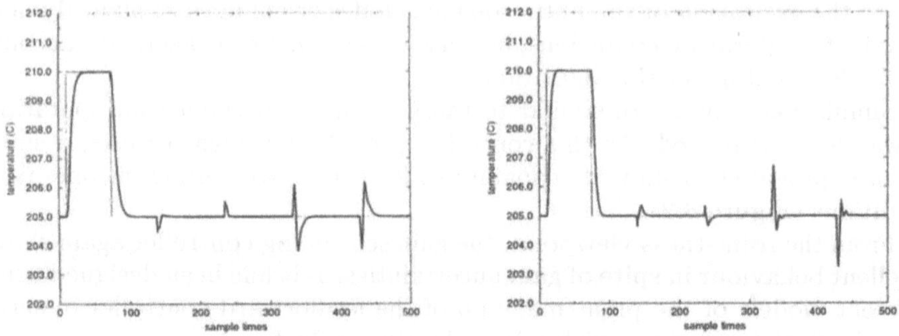

plant output with R=0.86 plant output with R=0.5

GAIN SCHEDULING GPC CONTROLLER. LOW FLOW CONDITIONS

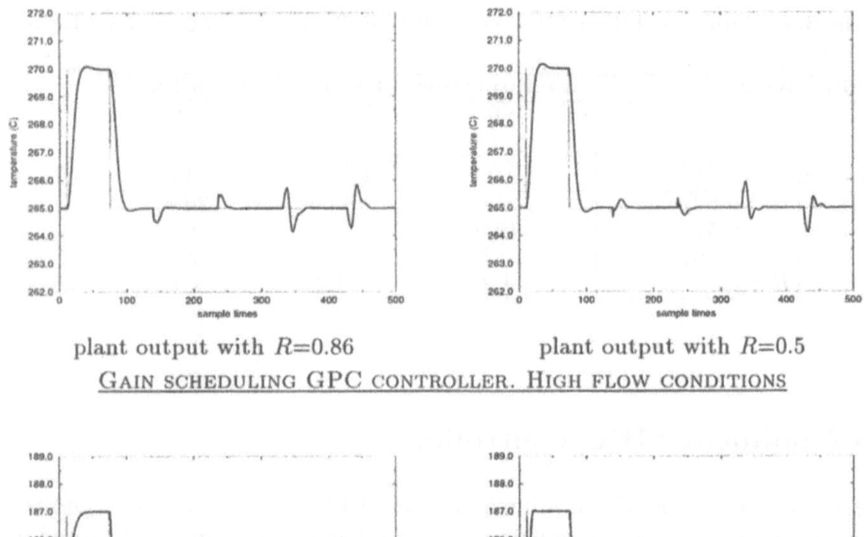

plant output with R=0.86 plant output with R=0.5

GAIN SCHEDULING GPC CONTROLLER. HIGH FLOW CONDITIONS

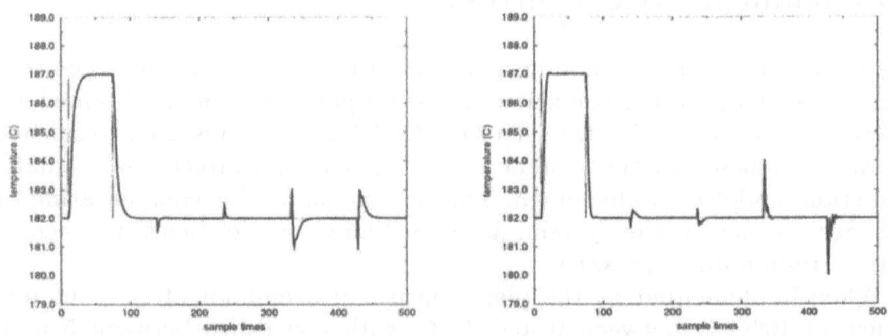

plant output with R=0.86 plant output with R=0.5

Fig. 8.2. Simulation with the gain scheduling GPC controller obtaining a nonlinear model of the plant may take considerable time which increases the mentioned commissioning time. As the nonlinear model used is complex, the computational effort of this controller is greater than that of other simple controllers. When implemented at the solar plant, each control signal calculation takes about 1 second, which is not a problem for the actual application.

Results achieved in the simulations with this control scheme are also excellent, as the inclusion of the nonlinear model in the prediction of the *free response* allows operation under the whole range of working conditions and disturbances to be explicitly accounted for. Differences arise from the plant results, as the overshoot and rise time obtained during different simulations are smaller than those obtained at the plant. This is a consequence of model inaccuracies, mainly from the dead time viewpoint.

With the disturbance profile used in the simulation, errors can arise due to the use of a constant value (equal to the actual one) of the inlet oil temperature and solar radiation in the predictions, as the controller is not able to predict future disturbances. Thus, when a step change is produced in the disturbances, because the previous predictions along the prediction horizon were carried out with another value of the disturbances, the reaction of the controller has to change this tendency. In the plant operation, results are usually better due to a less drastic profile of the disturbances. If the controller is to be designed to cope with hard disturbances, a trade–off can be found in the design stage of the controller by appropriately choosing the control and prediction horizons and the weighting factor.

In the case of the nonlinear GPC controller, the use of a wrong value of mirror reflectivity produces the disturbance predictions to be multiplied by a wrong scale factor. This fact affects the *free response* calculation, because the incremental formulation of the disturbance prediction is multiplied in order to be used by the nonlinear model by a wrong scale factor. This produces the calculation of the forced response to systematic errors that are corrected by the inclusion of integral action (CARIMA formulation). This was also observed in results obtained during real operation with both the CARMA and CARIMA models of the plant.

The values of the estimated reflectivity used by the feedforward controller which assure stable operation and performance criteria are shown in Table 8.5. As this controller provides quite a fast response to reference changes and disturbances in the nominal case ($R = 0.86$), when a value of $R = 0.5$ is used, the control action is quite aggressive, producing fast changes in manipulable and output variables, and large values of the CE criterion.

8.7 Frequency domain adaptive IMC controller

This strategy seems to be adequate to achieve fast set point tracking characteristics without oscillations caused by the excitation of the antiresonance

Table 8.5. Nonlinear GPC controller: Reflectivity range and performance indexes

Reflectivity		**medium flow**	**low flow**	**high flow**
R_{min}		≥ 0.2	≥ 0.2	≥ 0.2
R_{max}		≤ 1.0	≤ 1.0	≤ 1.0
Performance indexes		$(\times 10^3)$	$(\times 10^3)$	$(\times 10^3)$
ITAE	$(R = 0.86)$	6.8	13.2	7.6
ITAE	$(R = 0.5)$	34.6	56.3	30.1
CE	$(R = 0.86)$	0.7	1.3	0.5
CE	$(R = 0.5)$	5.7	13.6	1.35

modes. Nevertheless, the effect of disturbances acting on the system and the use of non-ideal filters with a small number of interpolating points in the controller design can lead the simple identification mechanisms to provide wrong values of the estimated parameters at each frequency band, and even to steady state errors. This is a critical fact when bands corresponding to zero frequency and the one near the first antiresonance mode are being identified. Although the use of bandpass filters allows for disturbances whose main components correspond to bands not chosen in the design to be rejected, different exogenous signals can cause the values of the estimated parameters to move away from the correct ones. To avoid these problems, different supervisory mechanisms have been introduced, obtaining overshoots of about 0 to 2 ^oC in the real plant operation after set point changes of 10°C, with rise times between 8 and 11 minutes and settling times between 8 and 20 minutes. Acceptable disturbance rejection capabilities are obtained with this scheme. This controller can be easily implemented once equations have been obtained. The computational effort is small (negligible) in the actual application, but will increase if the number of interpolating points used for control purposes is augmented.

Analysis of the results obtained both in real operation and simulations indicates that only acceptable behaviour can be found with this control scheme. In systems with resonance characteristics, such as the one dealt with in this book, the number of selected points in the Nyquist diagram of the system at the design stage of the algorithm must be increased if better performance is to be achieved.

The controller design has been performed by using only three points of the Nyquist diagram to show an easy implementable algorithm, with very short commissioning time and small computational effort. In this way, the lower antiresonance peak can be well located under certain operating conditions, but a wrong estimation can be produced if this frequency point changes.

This problem can be overcomed by choosing more points at the design stage of the algorithm, in order to diminish the uncertainty about the location and frequency characteristics of the antiresonance peaks. This improvement leads to an increase in the number of simple cascade estimators that have to be used.

If a wrong estimation of the steady state gain is made, this scheme can lead to steady state errors, which make the ITAE criterion achieve high values. This criterion has been used taking into account possible small steady state errors, as is done with the robust adaptive GPC controller.

It is to be expected adaptation mechanism to be able to cope with steady state gain variations, as the plant being identified is constituted of the feedforward controller in series with the field. In this case, performance deterioration is mainly due to changes in the delay of the field due to operation under low flow conditions.

The values of the estimated reflectivity used by the feedforward controller which assure a stable operation and those of the performance criteria are included in Table 8.6.

Table 8.6. Adaptive IMC controller: Reflectivity range and performance indexes

Reflectivity	**medium flow**	**low flow**	**high flow**
R_{min}	≥ 0.3	≥ 0.3	≥ 0.2
R_{max}	≤ 1.0	≤ 1.0	≤ 1.0
Performance indexes	$(\times 10^3)$	$(\times 10^3)$	$(\times 10^3)$
ITAE $(R = 0.86)$	36.2	97.1	42.4
ITAE $(R = 0.5)$	43.5	169.3	41.5
CE $(R = 0.86)$	1.1	2.5	1.2
CE $(R = 0.5)$	4.4	3.6	2.1

8.8 Robust LQG/LTR controller

Acceptable set point tracking characteristics are obtained throughout the whole range of operating conditions (but slower than other developed schema), with null step response steady state error (integral effect has been included during the design stage). The use of a simple linear model causes the modelling uncertainties to be high. The disturbance rejection capabilities are also expected to be acceptable, as this scheme allows for having a small sensitivity function at low frequencies.

When operating at the plant, the overshoot after a set point change of 10°C has been observed to be about 0 – 3 ^{o}C, with rise times of between 8 and 11 minutes and settling times of between 8 and 11 minutes.

The behaviour under solar radiation or inlet oil temperature disturbances is acceptable if a trade–off is found between good set point tracking characteristics and disturbance rejection. The computational effort is negligible and the commissioning time is also small as the design stage can easily be performed using standard CAD tools.

With this controller similar results as those obtained with the fixed PID controller are to be expected, but the design of the controller has been performed from a robust viewpoint to limit the deterioration produced by operating point changes and unmodelled dynamics (theoretically, the robustness increase in the design introduced by the loop transfer recovery mechanism should produce a decrease in variations of the system response in spite of the mentioned factors). It is evident that the PID controller design could be detuned to obtain slower response with better robustness characteristics. Nevertheless, this is not a clear matter, as there is not a direct relationship between controller detuning and the achieved robustness. In the case of the LQG/LTR controller, the design allows the control engineer to look for these desired robustness characteristics.

Results obtained in the simulations show how robustness also deteriorates at low flow conditions with this controller, but control system performance characteristics do not suffer from accentuated changes when different operating conditions are considered. As can be appreciated from data in Table 8.7, ITAE criterion changes at low flow conditions for different values of R are not quite as large as those obtained with other fixed controllers. Notice how the control effort is also adequate, showing that oscillations both in control and controlled signals are avoided.

Results are conservative but acceptable, as the use of a simple fixed first order model (with a fixed zero) in the design, corresponding to medium flow conditions (8 l/s) makes the uncertainty under low flow conditions be high. Although a small sensitivity to this kind of uncertainty was to be expected, the use of such a simple model limits the controller performance, mainly under low frequency (steady state gain) uncertainty, as occurs when wrong values of mirror reflectivity are used. Notice that robust control theory usually works under the assumption of limited values of low frequency uncertainty, which augments as frequency increases. On the other hand, the recovery degree has been adjusted to provide a trade–off between disturbance rejection capabilities and set point tracking, leading to slow responses to set point changes if the robustness degree is increased.

The values of the estimated reflectivity used by the feedforward controller which assure a stable operation are also included in Table 8.7:

Table 8.7. LQG/LTR controller: Reflectivity range and performance indexes

Reflectivity	**medium flow**	**low flow**	**high flow**
R_{min}	≥ 0.2	≥ 0.2	≥ 0.2
R_{max}	≤ 1.0	≤ 1.0	≤ 1.0
Performance indexes	$(\times 10^3)$	$(\times 10^3)$	$(\times 10^3)$
ITAE ($R = 0.86$)	33.2	92.7	37.8
ITAE ($R = 0.5$)	31.8	103.2	38.3
CE ($R = 0.86$)	1.1	2.3	1.1
CE ($R = 0.5$)	1.6	7.2	0.7

8.9 Heuristic incremental fuzzy PI controller (IFPIC)

The use of a PI structure for the fuzzy logic controller is restrictive in some sense, as fuzzy logic on its own is capable of achieving the performance required, as shown when designing the FLC. As has been mentioned, the intention of using this controller was to see how the performance of classical PI controllers can be improved by a simple fuzzy logic approach. In this way, the expected performance is that of an enhanced PI controller, which is adapted to cope with different situations in the time response of the system.

Observations from plant results indicate that the overshoot obtained after a set point change of 10°C is between 2 and 4 °C depending on the operating point, with rise times between 8 and 11 minutes and settling times between 8 and 18 minutes. Acceptable disturbance rejection capabilities are also obtained. The computational effort in this case is also negligible, as the inference mechanism used in the design of the IFPIC is quite simple. The commissioning time of this controller is small if there is an accumulated experience in operating with PID controllers at the plant.

Fuzzy logic controllers are affected by the wrong use of mirror reflectivity in the feedforward controller, but can compensate for these effects by the control action provided by the nonlinear control surface. The performance under low flow conditions deteriorates due to the same reasons as those commented on in previous sections.

When performing the simulations results obtained also show very similar characteristics to those seen in real operation. The values of the estimated reflectivity used by the feedforward controller which assure stable operation have been obtained as those shown in Table 8.8.

Table 8.8. IFPIC controller: Reflectivity range and performance indexes

Reflectivity	**medium flow**	**low flow**	**high flow**
R_{min}	≥ 0.3	≥ 0.3	≥ 0.2
R_{max}	≤ 1.0	≤ 1.0	≤ 1.0
Performance indexes	$(\times 10^3)$	$(\times 10^3)$	$(\times 10^3)$
ITAE $(R = 0.86)$	29.8	88.9	28.4
ITAE $(R = 0.5)$	26.0	122.5	19.1
CE $(R = 0.86)$	1.2	2.3	1.2
CE $(R = 0.5)$	1.2	8.4	4.7

8.10 Heuristic fuzzy logic controller (FLC)

The design features of this kind of controller and its nonlinear nature indicate that good set point tracking characteristics can be achieved with null or small steady state error if adequate rules are introduced in the design stage. Oscillations due to antiresonance modes can appear if a fast response is required, because these significant frequency domain characteristics are not taken into account. Poor behaviour in spite of disturbances acting on the system was expected, as this constitutes a low gain controller when operating very close to the set point. The inclusion of the feedforward controller in series with the plant attenuates the effects of the inlet oil temperature and solar radiation changes. Deviations not compensated for by the feedforward controller are corrected by the control signal provided by the FLC controller, but in a less efficient way than other type of controllers with high gain around the operating point. If the slope of the controller surface changes around the central zone, better results can be achieved in spite of disturbances acting on the system, but an underdamped response is obtained if set point changes are produced.

Values of the overshoot after a set point change of 10°C have been measured to be within the range 0 - 2 °C depending on the operating point, with rise times of between 7 and 10 minutes and settling times of between 7 and 16 minutes. Acceptable disturbance rejection capabilities are also obtained (better than those expected from theoretical considerations).

The large experience achieved during the course of several years has provided a very convenient fuzzy logic controller design, which appropriately implemented allows very good results to be obtained in spite of set point changes. The outlet oil temperature profile changes with the operating point as the oil flow has not been taken into account at the design stage of the algorithm, but very acceptable results can be obtained throughout the whole range of operating conditions. This design can be improved by incorporating the oil flow signal as an input to the table, but the controller design becomes more complex and less "graphical".

The values of the estimated reflectivity used by the feedforward controller which assure stable operation and performance criteria are included in Table 8.9.

Table 8.9. FLC controller: Reflectivity range and performance indexes

Reflectivity	**medium flow**	**low flow**	**high flow**
R_{min}	≥ 0.3	≥ 0.3	≥ 0.2
R_{max}	≤ 1.0	≤ 1.0	≤ 1.0
Performance indexes	$(\times 10^3)$	$(\times 10^3)$	$(\times 10^3)$
ITAE $(R = 0.86)$	28.8	86.2	19.4
ITAE $(R = 0.5)$	16.0	97.6	8.6
CE $(R = 0.86)$	1.1	2.3	1.0
CE $(R = 0.5)$	1.1	8.1	0.8

8.11 Conclusions

As expected, all the developed controllers have a set of advantages and drawbacks. The selection of one control strategy has to be based not only on the optimization of a determined behaviour given by classical performance indexes such as rise time, overshoot, etc., but also taking into account other kinds of considerations, such as having a limited control effort to avoid pump deterioration, good disturbance rejection capabilities, small overshoot at the starting phase of the operation, adequate commissioning time, etc. Other factors can also be taken into account at the design stage of the control scheme, such as easiness of implementing the controller specifications, the necessity of a simple, extended and even nonlinear model of the plant, the uncertainty of such kinds of models, etc. All these factors make the selection of an "ideal" controller for distributed solar collector fields be an open matter, depending on the particular characteristics of each plant (operating temperatures, tube lengths, experience of the operators, etc.). In general it can be said that if the plant model is known, model predictive control either with a gain scheduling linear model or a nonlinear model gives the best results. Adaptive control has some problems for this particular type of plant as the parameter changes occur in a similar time scale as process dynamics, and supervisory heuristic rules have to be added to make it work. If the plant model is not known, fuzzy logic control or robust control design techniques seem to be the best approach.

Different advanced control strategies have been analyzed throughout the book and applied to the control of the distributed solar collector field of the Plataforma Solar de Almería. This plant constitutes an interesting system to demonstrate how advanced controllers can be implemented in industrial practice. We hope this contribution will constitute a step towards the decreasing of the existing gap between theory and practice.

REFERENCES

1. A.Y. Allidina and F.M. Hughes. Generalized self-tuning controller with pole assignment. *Proc. IEE*, 127-D:13–18, 1980.
2. J.C. Allwright. *Advances in Model-Based Predictive Control, chapter: On min-max Model-Based Predictive Control.* Ed. D.W. Clarke, Oxford University Press, 1994.
3. B.D.O. Anderson. From Youla-Kucera to Identification, Adaptive and Nonlinear Control. In *13th Triennial World Congress (IFAC), San Francisco, USA*, volume Plenary Session, pages 39–59, 1996.
4. K.J. Aström. The future of control. *Modeling, Identification and Control. A Norwegian Research Bulletin*, 15(3):127–134, 1994.
5. K.J. Aström and B. Wittenmark. *Computer Controlled Systems. Theory and Design.* Prentice-Hall, 1984.
6. K.J. Aström and B. Wittenmark. *Adaptive Control.* Addison-Wesley, 1989.
7. G.J. Balas, J.C. Doyle, K. Glover, A. Packard, and R. Smith. *μ Analysis and Synthesis Toolbox.* Musyn Inc. and The MathWorks, Inc., 1991.
8. B.R. Barmish and H.I. Kang. A Survey of Extreme Point Results for Robustness of Control Systems. *Automatica*, 29(1):13–35, 1993.
9. M.S. Bazaraa and C.M. Shetty. *Nonlinear Programming.* Wiley, 1979.
10. M. Berenguel. *Contribuciones al Control de Colectores Solares Distribuidos.* PhD thesis, Universidad de Sevilla, 1996.
11. M. Berenguel and E.F. Camacho. Frequency Based Adaptive Control of Systems with Antiresonance Modes. In *Proc. of the 5th IFAC Symp. of Adaptive Systems in Control and Signal Processing (ACASP'95)*, pages 195–200, 1995.
12. M. Berenguel and E.F. Camacho. Frequency Based Adaptive Control of Systems with Antiresonance Modes. *Control Engineering Practice*, 4(5):677–684, 1996.
13. M. Berenguel, E.F. Camacho, and F.R. Rubio. *Operación con el Programa de Control del Campo ACUREX.* Plataforma Solar de Almería. Boletín interno, 1994.
14. M. Berenguel, E.F. Camacho, and F.R. Rubio. *Simulation Software Package for the Acurex Field.* Departamento de Ingeniería de Sistemas y Automática, ESI Sevilla (Spain), Internal Report, 1994.
15. L.T. Biegler and J.B. Rawlings. *Proc. Chemical Process Control IV, chapter: Optimization approaches to model predictive control.* Ed. Y. Arkun and W.H. Ray (AIChE, NY), 1991.
16. G. Bierman. *Factorization Methods for Discrete Estimation.* Academic Press, New York, 1977.
17. E.F. Camacho. Constrained Generalized Predictive Control. *IEEE Trans. on Automatic Control*, 38(2):327–332, 1993.

18. E.F. Camacho and M. Berenguel. *Advances in Model-Based Predictive Control, chapter: Application of Generalized Predictive Control to a Solar Power Plant.* Ed. D.W. Clarke, Oxford University Press, 1994.
19. E.F. Camacho and M. Berenguel. Application of Generalized Predictive Control to a Solar Power Plant. In *Proc. of The Third IEEE Conference on Control Applications, Glasgow, UK*, pages 1657–1662, 1994.
20. E.F. Camacho and M. Berenguel. Robust Adaptive Model Predictive Control of a Solar Plant with Bounded Uncertainties. *International Journal of Adaptive Control and Signal Processing*, March, 1997.
21. E.F. Camacho, M. Berenguel, and C. Bordóns. Adaptive Generalized Predictive Control of a Distributed Collector Field. *IEEE Trans. on Control Systems Technology*, 2(4):462–467, 1994.
22. E.F. Camacho, M. Berenguel, and F.R. Rubio. Application of a Gain Scheduling Generalized Predictive Controller to a Solar Power Plant. *Control Engineering Practice*, 2(2):227–238, 1994.
23. E.F. Camacho and C. Bordóns. Simple Implementation of Generalized Predictive Self-Tuning Controllers for Industrial Processes. *International Journal of Adaptive Control and Signal Processing*, 7:63–73, 1993.
24. E.F. Camacho and C. Bordóns. *Model Predictive Control in the Process Industry.* Springer-Verlag, 1995.
25. E.F. Camacho, F.R. Rubio, and J.A. Gutiérrez. Modelling and Simulation of a Solar Power Plant with a Distributed Collector System. In *Int. IFAC Symp. on Power Systems Modelling and Control Applications. Brussels*, 1988.
26. E.F. Camacho, F.R. Rubio, and F.M. Hughes. Self-tuning Control of a Solar Power Plant with a Distributed Collector Field. *IEEE Control Systems Magazine*, pages 72–78, 1992.
27. E.F. Camacho and Y. Tang. Adaptive Internal Model Control: a Frequency Domain Approach. In *Proc. of the ACASP'92 Symp.*, pages 27–32, 1992.
28. P.J. Campo and M. Morari. Robust Model Predictive Control. In *American Control Conference, Minneapolis, Minnesota*, 1987.
29. R. Carmona. *Análisis, modelado y control de un campo de colectores solares distribuidos con Sistema de seguimiento en un eje.* PhD thesis, Universidad de Sevilla, 1985.
30. M.A. Castro. *Simulación de Centrales de Energía Solar. Aplicación a la gestión energética.* PhD thesis, ESII Madrid, 1988.
31. D.W. Clarke. Application of Generalized Predictive Control to Industrial Processes. *IEEE Control Systems Magazine*, 122:49–55, 1988.
32. D.W. Clarke. Adaptive Predictive Control. In *Proc. Adaptive Systems in Control and Signal Processing (ACASP'95)*, pages 43–54, 1995.
33. D.W. Clarke and P.J. Gawthrop. Self-tuning Controller. *Proceedings IEE*, 122:929–939, 1975.
34. D.W. Clarke and C. Mohtadi. Properties of Generalized Predictive Control. *Automatica*, 25(6):859–875, 1989.
35. D.W. Clarke, C. Mohtadi, and P.S. Tuffs. Generalized Predictive Control - Part I. The Basic Algorithm. *Automatica*, 23(2):137–148, 1987.
36. D.W. Clarke, C. Mohtadi, and P.S. Tuffs. Generalized Predictive Control - Part II. Extensions and Interpretations. *Automatica*, 23(2):149–160, 1987.
37. D.W. Clarke and R. Scattolini. Constrained Receding-Horizon Predictive Control. *Proceedings IEE*, 138(4):347–354, 1991.
38. C.R. Cutler and B.C. Ramaker. Dynamic Matrix Control - A Computer Control Algorithm. In *Automatic Control Conference, San Francisco*, 1980.

39. R.M.C. De Keyser and A.R. Van Cuawenberghe. Extended Prediction Self-adaptive Control. In *IFAC Symp. on Identification and System Parameter Estimation, York, UK*, pages 1317–1322, 1985.
40. R.C. Dorf. *Modern Control Systems.* Addison-Wesley Publishing Company Inc., 1986.
41. J.C. Doyle, K. Glover, P.P. Khargonekar, and B.A. Francis. State-Space Solutions to Standard H_2 and H_∞ Control Problems. *IEEE Trans. on Automatic Control*, AC-34(8):831–846, 1989.
42. J.C. Doyle and G. Stein. Robustness with Observers. *IEEE Trans. on Automatic Control*, AC-24(4):607–611, 1979.
43. J.C. Doyle and G. Stein. Multivariable Feedback Design: Concepts for a Classical/Modern Synthesis. *IEEE Trans. on Automatic Control*, AC-31(1):4–16, 1981.
44. R. Fletcher. *Practical Methods of Optimization.* John Wiley & sons, 1990.
45. T.R. Fortescue, L.S. Kershenbaum, and B.E. Ydstie. Implementation of Self-tuning Regulators with Variable Forgetting Factors. *Automatica*, 17(6):831–835, 1981.
46. G.F. Franklin and J.D. Powell. *Digital Control of Dynamic Systems.* Addison-Wesley, London, 1980.
47. C.E. García, D.M. Prett, and M. Morari. Model Predictive Control: Theory and Practice - a Survey. *Automatica*, 25(3):335–348, 1989.
48. J.C. Geromel and J.J. Da Cruz. On the robustness of optimal regulators for nonlinear discrete-time systems. *IEEE Trans. on Automatic Control*, 32:703–710, 1987.
49. G.C. Goodwin, S.F. Graebe, and W.S. Levine. Internal Model Control of Linear Systems with Saturating Actuators. In *Proceedings of the ECC'93*, pages 1072–1077, 1993.
50. G.C. Goodwing, D.Q. Mayne, and M.E. Salgado. Uncertainty, Information and Estimation. In *Prep. of IFAC ACASP'89, Glasgow, UK*, pages 39–47, 1989.
51. R. Gorez, D. Galardini, and K.Y. Zhu. Internal Model Control and Disturbance Observers. In *Proc. of the 30th Conference on Decision and Control, Brighton, UK*, pages 229–234, 1991.
52. M.J. Grimble. *Robust Industrial Control: Optimal Design Approach for Polynomial Systems.* Prentice Hall, 1994.
53. T. Hägglund and K.J. Aström. Industrial Adaptive Controllers Based on Frequency Response Techniques. *Automatica*, 27(4):599–609, 1991.
54. M.A. Henson and D.E. Seborg. An internal model control strategy for nonlinear systems. *AIChE J.*, 37:1065–1081, 1991.
55. J. Jiang. Optimal Gain Scheduling Controller for a Diesel Engine. *IEEE Control Systems Magazine*, pages 42–48, 1994.
56. P. Källén. *A Frequency Domain Approach to Adaptive Control.* PhD thesis, Department of Automatic Control, Lund Institute of Technology, Sweden, 1994.
57. P. Källén and B. Wittenmark. A Frequency Domain Adaptive Controller. In *Proc. of the 12th IFAC World Congress, Sydney, Australia*, pages 67–70, 1993.
58. R.E. Kalman. When Is a Linear Control System Optimal? *Journal of Basic Engineering. Trans. of the ASME*, pages 51–60, 1964.
59. A. Kalt. *Distributed Collector System Plant Construction Report.* IEA/SSPS Operating Agent DFVLR, Cologne, 1982.
60. M. Kickert and R.H. Van Nauta Lemke. Application of a fuzzy controller in warm water plant. *Automatica*, 12:301–308, 1976.

61. J.P. King and H.E Mamdani. *Fuzzy reasoning and its applications, chapter: The application of fuzzy control systems to industrial processes.* Ed. Mamdani, E.H. and B.R. Gaines. New York: Academic Press, 1981.
62. R.L. Kosut. Adaptive Uncertainty Modelling: On-line Robust Control Design. In *Proc. ACC, Minneapolis, USA*, 1987.
63. B. Kouvaritakis, J.A. Rossiter, and A.O.T. Chang. Stable Generalized Predictive Control: An algorithm with Guaranteed Stability. *Proceeding IEE, Part D*, 139(4):349–362, 1992.
64. C. Kravaris and J.C. Kantor. Geometric methods for nonlinear process control: 1. Background. *Ind. Eng. Chem. Res.*, 29:2295–2310, 1990.
65. A.G. Kutnetsov and D.W. Clarke. *Advances in Model-Based Predictive Control, chapter: Application of Constrained GPC for Improving Performance of Controlled Plants.* Ed. D.W. Clarke, Oxford University Press, 1994.
66. H. Kwakernaak. Minimax Frequency Domain Performance and Robustness Optimization of Linear Feedback Systems. *IEEE Trans. on Automatic Control*, AC-30(10):994–1004, 1985.
67. W.H. Kwon and A.E. Pearson. On feedback stabilization of time-varying discrete linear systems. *IEEE Trans. on Automatic Control*, 23:479–481, 1979.
68. R.O. LaMaire, L. Valavani, M. Athans, and G. Stein. A Frequency Domain Estimator for Use in Adaptive Control System. In *Proc. ACC, Minneapolis, USA*, 1987.
69. M.P. Larsen. *Fuzzy reasoning and its applications, chapter: Industrial application of fuzzy logic control.* Ed. Mamdani, E.H. y B.R. Gaines. New York: Academic Press, 1981.
70. C.L. Lawson and R.J. Hanson. *Solving least-squares problems.* Prentice Hall, Englewood Cliffs, New Jersey, 1974.
71. J.R. Layne and K.M. Pasino. Fuzzy Model Reference Learning Control for cargo ship steering. *IEEE Control Systems Magazine*, pages 23–34, December, 1993.
72. C.C. Lee. Fuzzy Logic in Control Systems: Fuzzy Logic Controller – Part I. *IEEE Trans. on Systems, Man and Cybernetics*, 20(2):404–418, 1990.
73. C.C. Lee. Fuzzy Logic in Control Systems: Fuzzy Logic Controller – Part II. *IEEE Trans. on Systems, Man and Cybernetics*, 20(2):419–435, 1990.
74. L. Ljung. *System Identification.* Prentice-Hall, Englewood Cliffs, New Jersey, 1987.
75. L. Ljung. System identification in a MIC perspective. *Modeling, Identification and Control. A Norwegian Research Bulletin*, 15(3):153–159, 1994.
76. A.G.J. MacFarlane. Return-difference and return-ratio matrices and their use in analysis and design of multivariable feedback control systems. *Proc. IEE*, 117(10):2037–2049, 1970.
77. J.M. Maciejowski. Asymptotic Recovery for Discrete-Time Systems. *IEEE Trans. on Automatic Control*, AC-30(6):602–605, 1985.
78. A.P de Madrid, M. Santos, S. Dormido, and F. Morilla. *Advances in Model-Based Predictive Control, chapter: Constrained Generalized Predictive Control with Dynamic Programming.* Ed. D.W. Clarke, Oxford University Press, 1994.
79. E.H. Mamdani. Application of fuzzy algorithms for control of a simple dynamic plant. *Proceedings IEE*, 121:1585–1588, 1974.
80. J. Martín-Sánchez. Adaptive control for time-variant processes. *Int. Journal of Control*, 44(2):315–329, 1986.
81. A. Meaburn and F.M. Hughes. Resonance characteristics of distributed solar collector fields. *Solar Energy*, 51(3):215–221, 1993.

82. H. Messaoud, G. Favier, and R. Santos Mendes. Adaptive robust pole placement by connecting robust identification and control. In *Proc. IFAC Symp. on Adaptive Systems in Control and Signal Processing ACASP'92*, 1992.
83. R.H. Middelton. Frequency Domain Adaptive Control. In *Proc. IFAC Robust Adaptive Control, Newcastle, Australia*, pages 219–224, 1988.
84. S.H. Mo and J.P. Norton. Fast and robust algorithm to compute exact polytop parameter bounds. *Mathematics and Computers in simulation*, 32:481–493, 1990.
85. M. Morari. *Advances in Model-Based Predictive Control, chapter: Model Predictive Control: Multivariable Control Technique of Choice in the 1990s?* Ed. D.W. Clarke, Oxford University Press, 1994.
86. M. Morari and E. Zafiriou. *Robust Process Control.* Prentice-Hall, 1989.
87. E. Mosca. *Optimal, Predictive and Adaptive Control.* Prentice Hall, 1995.
88. E. Mosca, J.M. Lemos, and J. Zhang. Stabilizing I/O Receding Horizon Control. In *IEEE Conference on Decision and Control*, 1990.
89. E. Mosca and J. Zhang. Stable Redesign of Predictive Control. *Automatica*, 28:1229–1233, 1992.
90. G. de Nicolao and R. Scattolini. *Advances in Model-Based Predictive Control, chapter: Stability and Output Terminal Constraints in Predictive Control.* Ed. D.W. Clarke, Oxford University Press, 1994.
91. N.M.C. de Oliveira and L.T. Biegler. Constraint Handling and Stability Properties of Model-Predictive Control. *AIChE Journal*, 40(7), 1994.
92. V. Peterka. Predictor-based Self-tuning Control. *Automatica*, 20(1):39–50, 1984.
93. C.L. Phillips and H.T. Nagle Jr. *Digital Control Systems. Analysis and Design.* Prentice-Hall, 1984.
94. J. Rawlings and K. Muske. The Stability of Constrained Receding Horizon Control. *IEEE Trans. on Automatic Control*, 38:1512–1516, 1993.
95. J. Richalet. Industrial Applications of Model Based Predictive Control. *Automatica*, 29(5):1251–1274, 1993.
96. J. Richalet, S. Abu el Ata-Doss, C. Arber, H.B. Kuntze, A. Jacubash, and W. Schill. Predictive Functional Control. Application to Fast and Accurate Robots. In *Proc. 10th IFAC Congress, Munich*, 1987.
97. J. Richalet, A. Rault, J.L. Testud, and J. Papon. Algorithmic Control of Industrial Processes. In *4th IFAC Symposium on Identification and System Parameter Estimation. Tbilisi, URSS*, 1976.
98. J. Richalet, A. Rault, J.L. Testud, and J. Papon. Model Predictive Heuristic Control: Application to Industrial Processes. *Automatica*, 14(2):413–428, 1978.
99. F.R. Rubio. *Control Adaptativo de Procesos Industriales. Aplicación a una Central Solar.* PhD thesis, Universidad de Sevilla, 1985.
100. F.R. Rubio and J. Aracil. De vuelta al Diagrama de Bode pasando por el Control Optimo. *Informática y Automática*, 25(2):25–46, 1992.
101. F.R. Rubio, M. Berenguel, and E.F Camacho. Fuzzy Logic Control of a Solar Power Plant. *IEEE Trans. on Fuzzy Systems*, 3(4):459–468, 1995.
102. F.R. Rubio, F. Gordillo, and M. Berenguel. LQG/LTR Control of the Distributed Collector Field of a Solar Power Plant. In *Proc. of the 6th Int. IFAC Symp., San Francisco, USA*, 1996.
103. F.R. Rubio, F.M. Hughes, and E.F. Camacho. Self-tuning PI control of a Solar Power Plant. In *Preprints of IFAC Symposium in Adaptive Systems in Control and Signal Processing (ACASP'89), Glasgow, UK*, pages 335–340, 1989.
104. F.R. Rubio and M.J. López. *Control Adaptativo y Robusto.* Servicio de Publicaciones de la Universidad de Sevilla, 1996.

105. V. Ruiz and C. Gómez. *Centrales Eléctricas Termosolares. Energías Renovables y Medio Ambiente.* Ed. Centro de Ordenación del Territorio y Medio Ambiente del Ministerio de Obras Públicas y Urbanismo. Madrid, 1982.
106. M. Safonof, A.J. Laub, and G.L. Hartmann. Feedback Properties of Multivariable Systems: The Role and Use of the Return Difference Matrix. *IEEE Trans. on Automatic Control*, AC-26(1):47–65, 1981.
107. E.M. Scharf. Fuzzy logic could redefine robot control. *Robotica*, Feb.:11–13, 1985.
108. E.M. Scharf, N.J. Mondie, and E.H. Mamdani. *A Self-organizing algorithm for the control of a robot arm.* Queen Mary College University of London, Internal Report, 1983.
109. P. Scokaert and D.W. Clarke. *Advances in Model-Based Predictive Control, chapter: Stability and Feasibility in constrained predictive control.* Ed. D.W. Clarke, Oxford University Press, 1994.
110. D.E. Seborg. A perspective on advanced strategies for process control. *Modeling, Identification and Control. A Norwegian Research Bulletin*, 15(3):179–189, 1994.
111. B. Shahian and M. Hassul. *Control System Design using MATLAB.* Prentice Hall, 1993.
112. J.S. Shamma and M. Athans. Analysis of Gain-Scheduled Control for Nonlinear Plants. *IEEE Trans. on Automatic Control*, 35(8), 1990.
113. R.A. Shoureshi, K.B. Garg, D. Garg, J. Maulbetsch, J. Albus, B. Widrow, T. Samad, P. Antsaklis, A. Meystel, and K. Furuta. Intelligent Control Systems. In *13th IFAC World Congress, San Francisco, USA*, pages 101–106, 1996.
114. G. Stein and M Athans. The LQR/LTR Procedure for Multivariable Feedback Control Design. *IEEE Trans. on Automatic Control*, AC-32(2), 1987.
115. T. Sudkamp and R.J. Hammell II. Interpolation, Completion, and Learning Fuzzy Rules. *IEEE Trans. on Systems, Man, and Cybernetics*, 24(2):332–342, 1994.
116. Y. Tang. A Frequency Domain Approach to Robust Adaptive Control. In *Proc. ACC, Pittsburgh, USA*, 1989.
117. Y. Tang, E.F. Camacho, and J. Flores. Frequency Domain Control: Band-wise Compensation. *Automatica*, 31(5):735–740, 1995.
118. Y. Tang and R. Ortega. Adaptive Tuning to Frequency Response Identification. In *Proc. of the IFAC World Congress, Tallin (URSS)*, pages 271–276, 1990.
119. T.T.C. Tsang and D.W. Clarke. Generalised Predictive Control with Input Constraints. *IEE Proceedings*, 135 Pt. D(6):451–460, 1988.
120. S. Tzafestas and N.P. Papanikolopoulos. Incremental Fuzzy Expert PID Control. *IEEE Trans. on Industrial Electronics*, 37(5):365–371, 1990.
121. S.M. Veres and J.P. Norton. Predictive Self-Tuning Control by Parameter Bounding and Worst-case Design. *Automatica*, 29(4):911–928, 1993.
122. P.E. Wellstead, D. Prager, and P. Zanker. A pole assignment self-tuning regulator. *Proc. IEE*, 126-128:781–787, 1978.
123. S. Yasunobu, S. Miyamoto, T. Takaoka, and H. Ohshiroa. Application of predictive fuzzy control of automatic train operation controller. In *Proc. IECON'84*, 1984.
124. B.E. Ydstie. Extended Horizon Adaptive Control. In *Proc. 9th IFAC World Congress, Budapest, Hungría*, 1984.
125. E. Zafiriou. Robust Model Predictive Control of Processes with hard Constraints. *Computers in Chemical Engineering*, 14(4/5):359–371, 1990.

126. E Zafiriou. On the Robustness of Model Predictive Controllers. In *4th Int. Conf. on Chemical Process Control, South Padre Island, Texas*, 1991.
127. E. Zarza. *Solar Thermal Desalination Project Phase II. Results and Final Project Report.* Colección Documentos Ciemat, 1995.
128. J.G. Ziegler and N.B. Nichols. Optimum settings for automatic controllers. *Trans. ASME*, 64:759–768, 1942.

INDEX

MIX
Papier aus verantwortungsvollen Quellen
Paper from responsible sources
FSC® C105338

If you have any concerns about our products,
you can contact us on
ProductSafety@springernature.com

In case Publisher is established outside the EU,
the EU authorized representative is:
Springer Nature Customer Service Center GmbH
Europaplatz 3, 69115 Heidelberg, Germany

Printed by Libri Plureos GmbH
in Hamburg, Germany